Computers and Medicine

Bruce I. Blum, *Editor*

Computers and Medicine

Computer-Assisted Medical Decision Making

Volume 2

Edited by
James A. Reggia and Stanley Tuhrim

With 84 Illustrations

Springer-Verlag
New York Berlin Heidelberg Tokyo

James A. Reggia, M.D., Ph.D.
Department of Neurology
University of Maryland Hospital
Baltimore, Maryland 21201
U.S.A.

Stanley Tuhrim, M.D.
Department of Neurology
University of Maryland Hospital
Baltimore, Maryland 21201
U.S.A.

Series Editor
Bruce I. Blum
Applied Physics Laboratory
The Johns Hopkins University
Laurel, Maryland 20707
U.S.A.

Library of Congress Cataloging in Publication Data
Main entry under title:
Computer-assisted medical decision making.
    (Computers and medicine)
    Includes bibliographies and index.
    1. Medicine—Decision making—Data processing—
Addresses, essays, lectures.   2. Medicine, Clinical—
Decision making—Data processing—Addresses, essays,
lectures.   3. Diagnosis—Data processing—Addresses,
essays, lectures.   4. Medical logic—Addresses,
essays, lectures.   I. Reggia, James A.   II. Tuhrim, Stanley.
III. Series: Computers and medicine (New York, N.Y.)
[DNLM: 1. Computers. 2. Decision Making. 3. Diagnosis,
Computer Assisted. W 26.5 C7383]
R723.5.C64   1985   610'.28'54   85-2697

Typeset by University Graphics, Inc., Atlantic Highlands, New Jersey.

9 8 7 6 5 4 3 2 1
ISBN-13: 978-1-4612-9567-9   e-ISBN-13: 978-1-4612-5108-8
DOI: 10.1007/978-1-4612-5108-8

To the memory of Marty Epstein . . .

For the humanistic qualities he brought to the field
of computers in medicine

Series Preface

Computer technology has impacted the practice of medicine in dramatic ways. Imaging techniques provide noninvasive tools which alter the diagnostic process. Sophisticated monitoring equipment presents new levels of detail for both patient management and research. In most of these high technology applications, the computer is embedded in the device; its presence is transparent to the user.

There is also a growing number of applications in which the health care provider directly interacts with a computer. In many cases, these applications are limited to administrative functions, e.g., office practice management, location of hospital patients, appointments, and scheduling. Nevertheless, there also are instances of patient care functions such as results reporting, decision support, surveillance, and reminders.

This series, Computers and Medicine, will focus upon the direct use of information systems as it relates to the medical community. After twenty-five years of experimentation and experience, there are many tested applications which can be implemented economically using the current generation of computers. Moreover, the falling cost of computers suggests that there will be even more extensive use in the near future. Yet there is a gap between current practice and the state-of-the-art.

This lag in the diffusion of technology results from a combination of two factors. First, there are few sources designed to assist practitioners in learning what the new technology can do. Secondly, because the potential is not widely understood, there is a limited marketplace for some of the more advanced applications; this, in turn, limits commercial interest in the development of new products.

In the next decade, one can expect the field of medical information science to establish a better understanding of the role of computers in medicine. Furthermore, those entering the health care professions already will have had some formal training in computer science. For the near term, however, there is a clear need for books designed to illustrate how computers can assist in the practice of medicine. For without these collections, it will be very difficult for the practitioner to learn about a technology which certainly will alter his or her approach to medicine.

And that is the purpose of this series: the presentation of readings about the interaction of computers and medicine. The primary objectives are to describe the current state-of-the-art and to orient medical and health professionals and students with little or no experience with computer applications. We hope that this series will help in the rational transfer of computer technology to medical care.

Laurel, Maryland   BRUCE BLUM

Preface

A little knowledge that acts is worth
infinitely more than much knowledge that is
idle.

K. Gibran

Decision-making is ... something which
concerns all of us, both as makers of the
choice and as sufferers of the consequences.

D. Lindley

Who is going to use all that computer
power?

A. Robinson

The idea of computer programs that can directly assist the physician with
decision making is both intriguing and disconcerting. It is intriguing
because of the potential to improve medical care and medical education,
not so much through the discovery of new biomedical knowledge as
through more effective application of that which already exists. It is also
disconcerting because of the potential for abuse and alteration to the
practice of contemporary medicine.

For over a quarter of a century, research has been underway to develop
new methods for computer-aided decision making in medicine, and to
evaluate critically these methods in clinical practice. The objective of this
book and its companion volume is to provide a collection of articles that
accurately reflects the achievements of this past effort. In selecting material for these volumes we attempted to choose representative examples of
technological developments which placed the evolution of computer-
assisted medical decision making in a historical perspective. We also
tried to provide a glimpse at the future directions the field may be taking.
Thus, the sequence of articles in these volumes progresses from the
more mature algorithmic and statistically oriented approaches which
pioneered the field to the more recent and unsettled technology of artificial intelligence. While we obviously could not include many fine pieces
of past work in these volumes, we believe that the articles which are

included provide both an excellent reference collection and a solid introduction to the field.

Editing a collection like this requires the active support of a number
of people. We would like to express our thanks to the authors for permission to use the papers in these two volumes, and to the people at
Springer-Verlag for converting the papers into high quality books. This
collection would never have been completed without the technical assistance of Gloria Kimbles, who served as our editorial assistant. We also
wish to acknowledge the constructive suggestions and guidance provided
to us by Bruce Blum. Finally, we want to thank Carol and Betty for their
support and patience while we undertook Yet Another Project.

Baltimore, Maryland   JIM REGGIA
                                                                  STAN TUHRIM

Reader's Guide

This is the second of two volumes that form a collection of articles on Computer-Assisted Medical Decision making (CMD) systems. The first volume concentrated on algorithmic and statistical approaches to building CMD systems. This second volume focuses on artificial intelligence (AI) techniques and general isssues related to building CMD systems.

The first section in this volume, **Rule-Based Systems**, focuses on one of the most widely used AI techniques: rule-based inference. The papers in this section present two of the earliest and best known examples of rule-based CMD systems, MYCIN (Davis) and CASNET/Glaucoma (Kulikowski). Additional papers discuss the difficulties encountered in representing knowledge in a rule format (Reggia), provide an example of an operational system (Speedie), and discuss risk analysis and critiquing of physician management plans using a rule-based approach (Miller).

The second section in this volume, **Cognitive Models**, presents four articles on diagnostic reasoning and its machine emulation. The first of these papers describes the results of an empirical study of diagnostic reasoning (Kassirer). This paper characterizes such reasoning as a hypothesize-and-test process and provides a wealth of information about the details involved. Subsequent articles describe three models of the diagnostic reasoning process previously implemented as CMD systems: PIP (Pauker), INTERNIST (Miller), and parsimonious set covering (Reggia).

The final section, **Related Issues**, focuses on several important topics relevant to developing CMD systems. Measuring the accuracy of probabilistic predictions is discussed (Shapiro), and three general purpose "CMD system generators" are described (Kulikowski; Reggia; Horn). In addition to these general frameworks for acquiring knowledge, an approach to the automatic discovery of new knowledge from clinical databases is presented (Blum). The section closes with two papers on the use of basic "domain principles" (Swartout; Patil).

Contents for Volume 2

IV. Rule-Based Systems

V. Cognitive Models

VI. Related Issues

Contributors to Volume 2

The following is a list of contributors to this volume. The contributed chapter number(s) is indicated in parentheses.

Robert Beardsley, Ph.D. (18)
Robert L. Blum (28)
Bruce Buchanan (15)
Walter Buchstaller (27)
Randall Davis (15)
G. Anthony Gorry, Ph.D. (20,21)
Werner Horn (27)
Jerome P. Kassirer, M.D. (20,21)
David A. Knapp, Ph.D. (18)
Casimir A. Kulikowski (16,25)
Perry L. Miller, M.D., Ph.D.(19)
Randolph A. Miller, M.D. (22)
Jack D. Myers, M.D. (22)
Dana S. Nau (23)
Jack H. Ostroff, M.D. (25)
Francis B. Palumbo, Ph.D. (18)
Ramesh S. Patil (30)

Stephen G. Pauker, M.D. (21)
Barry T. Perricone (26)
Harry E. Pople, Jr., Ph.D. (22)
Thomas R. Price, M.D. (26)
Thaddeus P. Pula (26)
James A. Reggia, M.D., Ph.D.
  (17,23,26)
William B. Schwartz, M.D. (21,30)
Alan R. Shapiro, M.D. (24)
Edward Shortliffe (15)
Stuart M. Speedie, Ph.D. (18)
William R. Swartout (29)
Peter Szolovits (30)
Robert Trappl (27)
Pearl Y. Wang (23)
Gio C.M. Wiederhold (28)

Contents for Volume 1

Contributors to Volume 1

The following is a list of contributors to Volume 1. The contributed chapter number(s) is indicated in parentheses.

G. Octo Barnett, M.D. (11)
Moshe Ben-Bassat, Ph.D. (12)
Howard L. Bleich, M.D. (3)
Jane Buell, B.A. (4)
Richard W. Carlson, M.D., Ph.D. (12)
John T. Carpenter, M.D. (5)
A. L. Dannenberg, M.D. (13)
Mark D. Davenport (12)
F.T. de Dombal, M.D., F.R.C.S. (7,8)
Alvin Essig, M.D. (14)
J.F. Fries (13)
Dennis G. Fryback (9)
G. Anthony Gorry, Ph.D. (11,14)
Jane C. Horrocks (7,8)
Roger W. Jelliffe, M.D. (4)
Robert Kalaba, Ph.D. (4)
Jerome P. Kassirer, M.D. (14)
Mohamed Latif (12)
D.J. Leaper, M.B., Ch.B. (7,8)
Robert S. Ledley (2)

Edward H. Lipnick (12)
Lee B. Lusted, M.D. (2)
A.P. McCann, M.Sc., Ph.D. (7)
Emmanuel Mesel, M.D. (5)
Larry D. Portigal, M.S. (12)
Venod K. Puri (12)
James A. Reggia, M.D., Ph.D. (1,10)
John A. Schriver, M.D. (12)
William B. Schwartz, M.D. (14)
A.R. Shapiro, M.D. (13)
Ronald Smith, M.D. (12)
J.R. Staniland (7,8)
Robert Stephenson, Ph.D. (6)
Alan F. Toronto, M.D. (6)
Stanley Tuhrim, M.D. (1)
L. George Veasey, M.D. (6)
Homer R. Warner, M.D., Ph.D. (6)
Max Harry Weil, M.D., Ph.D. (12)
David Wirtschafter, M.D. (5)
Ronald J. Zagoria, M.D. (10)

IV. RULE-BASED SYSTEMS

15

Production Rules as a Representation for a Knowledge-Based Consultation Program

Randall Davis, Bruce Buchanan, and
Edward Shortliffe

The MYCIN system has begun to exhibit a high level of performance as a consultant on the difficult task of selecting antibiotic therapy for bacteremia. This report discusses issues of representation and design for the system. We describe the basic task and document the constraints involved in the use of a program as a consultant. The control structure and knowledge representation of the system are examined in this light, and special attention is given to the impact of production rules as a representation. The extent of the domain independence of the approach is also examined.

1. Introduction

Two recent trends in artificial intelligence research have been applications of AI to "real-world" problems, and the incorporation in programs of large amounts of task-specific knowledge. The former is motivated in part by the belief that artificial problems may prove in the long run to be more a diversion than a base to build on, and in part by the belief that the field has developed sufficiently to provide techniques capable of tackling real problems.

The move toward what have been called "knowledge-based" systems represents a change from previous attempts at generalized problem solvers (as, for example, GPS). Earlier work on such systems demonstrated that while there was a large body of useful general purpose techniques (e.g., problem decomposition into subgoals, heuristic search in its many forms), these did not by themselves offer sufficient power for high performance. Rather than non-specific problem solving power, knowledge-based systems have emphasized both the accumulation of large amounts

Reprinted with permission from *Artificial Intelligence, 8,* pp. 15–45. Copyright 1977 by North-Holland Publishing Company.

of knowledge in a single domain, and the development of domain-specific techniques, in order to develop a high level of expertise.

There are numerous examples of systems embodying both trends, including efforts at symbolic manipulation of algebraic expressions [21], speech understanding [19], chemical inference [3], the creation of computer consultants as interactive advisors for various tasks [14,29], as well as several others.

In this paper we discuss issues of representation and design for one such knowledge-based application program—the MYCIN system developed over the past three years as an interdisciplinary project at Stanford University,[1] and discussed elsewhere [27–30]. Here we examine in particular how the implementation of various system capabilities is facilitated or inhibited by the use of *production rules* as a knowledge representation. In addition, the limits of applicability of this approach are investigated.

We begin with a review of features which were seen to be essential to any knowledge-based consultation system, and suggest how these imply specific program design criteria. We note also the additional challenges offered by the use of such a system in a medical domain. This is followed by an explanation of the system structure, and its fundamental assumptions. The bulk of the paper is then devoted to a report of our experience with the benefits and drawbacks of production rules as a knowledge representation for a high performance AI program.

2. System Goals

The MYCIN system was developed originally to provide consultative advice on diagnosis of and therapy for infectious diseases—in particular, bacterial infections in the blood.[2] From the start, the project has been shaped by several important constraints. The decision to construct a high

---

[1]The MYCIN system has been developed by the authors in collaboration with:

Drs Stanley Cohen, Stanton Axline, Frank Rhame, Robert Illa and Rudolpho Chavez-Pardo, all of whom provided medical expertise; William van Melle, who made extensive revisions to the system code for efficiency and to introduce new features; Carlisle Scott, who (with William Clancey) designed and implemented the expanded natural language question answering capabilities.

[2]We have recently begun investigating extending the system. The next medical domain will be the diagnosis and treatment of meningitis infections. This area is sufficiently different to be challenging, and yet similar enough to suggest that some of the automated procedures we have developed may be quite useful. The paper by van Melle [34] describes how an entirely different knowledge base—one concerned with a problem in automobile repair—was inserted into the system, producing a small but fully functional automobile consultant program.

To reflect this broadening of interests, the project has been renamed the Knowledge-Based Consultation Systems Project. MYCIN remains the name of the medical program.

performance AI program in the consultant model brought with it several demands. First, the program had to be *useful* if we expected to attract the interest and assistance of experts in the field. The task area was thus chosen partly because of a demonstrated need: for example, in a recent year one of every four people in the U.S. was given penicillin, and almost 90% of those prescriptions were unnecessary [16]. Problems such as these indicate the need for more (or more accessible) consultants to physicians selecting antimicrobial drugs. Usefulness also implies competence, consistently high performance, and ease of use. If advice is not reliable, or is difficult to obtain, the utility of the program is severely impaired.

A second constraint was the need to design the program to accommodate a *large and changing body of technical knowledge.* It has become clear that large amounts of task-specific knowledge are required for high performance, and that this knowledge base is subject to significant changes over time [3,13]. Our choice of a production rule representation was significantly influenced by such features of the knowledge base.

A third demand was for a system capable of handling an *interactive dialog,* and one which was not a "black box." This meant that it had to be capable of supplying coherent explanations of its results, rather than simply printing a collection of orders to the user. This was perhaps the major motivation for the selection of a symbolic reasoning paradigm, rather than one which, for example, relied totally on statistics. It meant also that the "flow" of dialog—the order of questions—should make sense to a physician and not be determined by programming considerations. Interactive dialog required, in addition, extensive human engineering features designed to make interaction simple for someone unaccustomed to computers.

The choice of a medical domain brought with it additional demands [28]. Speed, access and ease of use gained additional emphasis, since a physician's time is typically limited. The program also had to fill a need well-recognized by the clinicians who would actually use the system, since the lure of pure technology is usually insufficient. Finally, the program had to be designed with an emphasis on its supportive role as a tool for the physician, rather than as a replacement for his own reasoning process.

Any implementation selected had to meet all these demands. Predictably, some have been met more successfully than others, but all have been important factors in influencing the system's final design.

3. System Overview

3.1. The Task

The fundamental task is the selection of therapy for a patient with a bacterial infection. Consultative advice is often required in the hospital because the attending physician may not be an expert in infectious diseases, as for example, when a cardiology patient develops an infection after heart surgery. Time considerations compound the problem. A specimen (of blood, urine, etc.) drawn from a patient may show some evidence of bacterial growth within 12 hours, but 24–48 hours (or more) are required for positive identification. The physician must therefore often decide, in absence of complete information, whether or not to start treatment and what drugs to use if treatment is required. Both of these may be difficult questions.

The task will become clearer by reviewing the initial and final parts of a sample dialog with the MYCIN system, shown in Figures 1 and 2 (italicized comments at the right are for explanatory purposes, and are not part of the actual dialog). A great deal of attention has been paid to making the interaction easy and natural. Questions have been carefully worded, for example, so as to set up expectations of responses without making them as uninteresting as multiple choice questions. Currently the dialog is directed primarily by the system—patient information can be entered only by answering each question in turn.

There are several options available to make the interaction more informative. The user can, for instance, ask for a rephrasing of the question or a display of some (or all) of the acceptable responses. If a requested item of information is unavailable, he can respond with **UNKNOWN (U,** or **UNK** for short). The user can also modify any answer which he is unsure of by attaching a number indicating his degree of certainty (as, for example, in questions 13 and 15).

The main purpose of the system is to provide advice, and it does so via the *consultation program* (Figure 3). In addition to this performance program, there are three adjunct programs that increase system utility and flexibility: (a) a *question answering program* which handles simple natural language inquiries about either the specific consultation, or about the general knowledge base of the program; (b) an *explanation program* for giving reasons why the performance program has requested an item of information or has drawn a particular conclusion; and (c) a *knowledge acquisition program* for extending the knowledge base through a dialog with an expert.

The system is written in INTERLISP, runs on a DEC KI-10 with approximately 130K of nonshared code, and is fast enough to permit convenient real-time interaction. The consultation program (with the required data structures) requires approximately 50K, the explanation program 10K, the question answering program 15K, the rule acquisition program 20K, and the rules themselves approximately 8K (the remainder includes a dictionary, information on drug properties, and various system utility features).

As described below, each of these four interrelated programs meets one or more of the goals outlined above.

```
- - - - - - - - PATIENT-1 - - - - - - - -
```
(1) Patient's name: (first-last)
••FRED SMITH

{*background data*}

(2) Sex:
••MALE
(3) Age:
••55
(4) Have you been able to obtain positive cultures from a site at which Fred Smith has an infection?
••YES
```
- - - - - - - - INFECTION-1 - - - - - - -
```
(5) What is the infection?
★★PRIMARY-BACTEREMIA
(6) Please give the date and approximate time when signs or symptoms of the primary-bacteremia (INFECTION-1) first appeared. (mo/da/yr)
••MAY 5, 1975
The most recent positive culture associated with the primary-bacteremia (INFECTION-1) will be referred to as:
```
- - - - - - - - CULTURE-1 - - - - - - - -
```
(7) From what site was the specimen for CULTURE-1 taken?
••BLOOD
(8) Please give the date and time when this blood culture (CULTURE-1) was obtained. (mo/da/yr)
••MAY 9, 1975
The first significant organism from this blood culture (CULTURE-1) will be referred to as:
```
- - - - - - - - ORGANISM-1 - - - - - - - -
```
(9) Enter the identity of ORGANISM-1:
••UNKNOWN

{*typically the identity is not known and must be inferred*}

(10) Is ORGANISM-1 a rod or coccus (etc.):
••ROD
(11) The gram stain of ORGANISM-1:
••GRAMNEG
(12) Have there been POSITIVE cultures from sites which are NOT associated with infections about which you are seeking advice?
••NO
(13) Did ORGANISM-1 grow aerobically?
••YES (8)
(14) Did ORGANISM-1 grow anaerobically?
••YES
(15) What is the suspected portal of entry of ORGANISM-1 into this sterile site?
••GI (6)
. . .

{*more questions follow, to gather sufficient information to infer possible identities of the infecting organisms*}

Figure 1. Initial segment of a session with the consultation system. User responses are in **boldface** and follow the double asterisks. Italicized comments at the right are not part of the actual dialog.

My therapy recommendation will be based on the following probable infection(s) and potential causative organism(s):

INFECTION-1 is PRIMARY-BACTEREMIA

{possible identities}

⟨Item 1⟩ The identity of ORGANISM-1 may be PSEUDOMONAS-AERUGINOSA
⟨Item 2⟩ The identity of ORGANISM-1 may be KLEBSIELLA-PNEUMONIAE
⟨Item 3⟩ The identity of ORGANISM-1 may be E. COLI
⟨Item 4⟩ The identity of ORGANISM-1 may be BACTEROIDES-FRAGILIS
⟨Item 5⟩ The identity of ORGANISM-1 may be ENTEROBACTER
⟨Item 6⟩ The identity of ORGANISM-1 may be PROTEUS-NON-MIRABILIS

    (46) On a scale of 0 to 4, where higher numbers indicate increasing severity how would you rate Fred Smith's degree of sickness?
••3
    (47) Does Fred Smith have a clinically significant allergic reaction to any antimicrobial agent?
••NO
    (48) Fred Smith's weight in kilograms (or ⟨number⟩ POUNDS):
••70

{note that 2 drugs cover for 6 potential organisms}

[Rec 1] My preferred therapy recommendation is as follows:
  In order to cover for Items ⟨1 2 3 5 6⟩:
    Give: GENTAMICIN
    Dose: 119 mg (1.7 mg/kg) q8h IV [or IM] for 10 days
    Comments: Modify dose in renal failure
  In order to cover for Item ⟨4⟩:
    Give: CLINDAMYCIN
    Dose: 595 mg (8.5 mg/kg) q6h IV [or IM] for 14 days
    Comments: If diarrhea or other GI symptoms develop, patient should be evaluated for possible pseudomembranous colitis.

Do you wish to see the next choice therapy?

{alternative choices are available}

••Y

Which, if any, of the above drugs do you wish eliminated from consideration?
••CLINDAMYCIN

[Rec 2] Next best therapy recommendation:
  In order to cover for Items ⟨2 3 4 5 6⟩:
    Give: CHLORAMPHENICOL
    Dose: 1 gm (15 mg/kg) q6h IV for 14 days
    Comments: Monitor patient's white count
  In order to cover for Item ⟨1⟩:
    Give: GENTAMICIN
  Dose: 119 mg (1.7 mg/kg) q8h IV [or IM] for 10 days
    Comments: Modify dose in renal failure

Figure 2. Final segment of a consultation. Alternative choices for therapies are generated if requested.

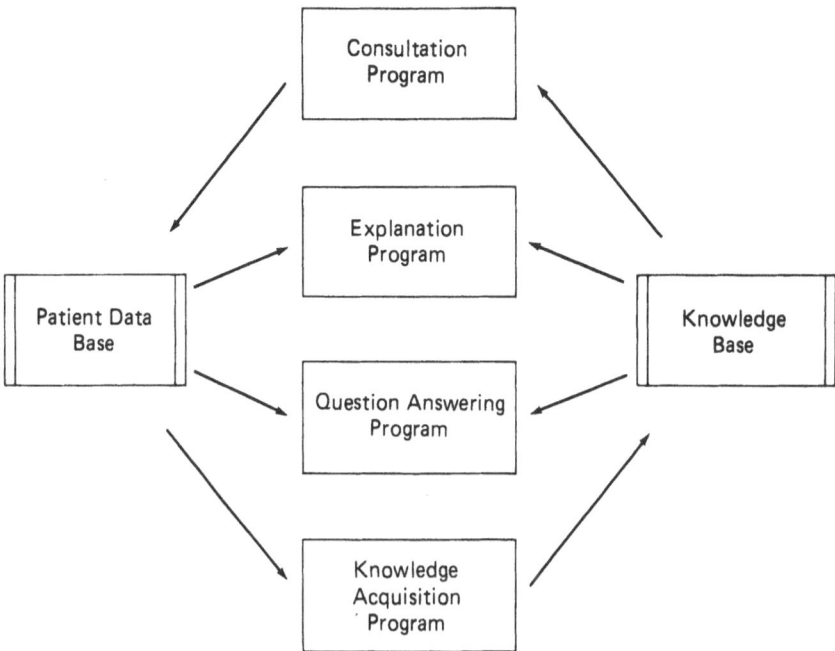

Figure 3. The six components of the system: four programs, the knowledge base, and the patient data base. All of the system's knowledge of infectious disease is contained within the knowledge base. Data about a specific patient collected during a consultation is stored in the patient data base. Arrows indicate the direction of information flow.

3.2. The Rules

The primary source of domain specific knowledge is a set of some 200 production rules, each with a premise and an action (Figure 4). The premise is a Boolean combination of predicate functions on associative triples. Thus each clause of a premise has the following four components:

⟨predicate function⟩ ⟨object⟩ ⟨attribute⟩ ⟨value⟩

There is a standardized set of 24 predicate functions (e.g., SAME, KNOWN, DEFINITE), some 80 attributes (e.g. IDENTITY, SITE, SENSITIVITY), and 11 objects (e.g. ORGANISM, CULTURE, DRUG), currently available for use as primitives in constructing rules. The premise is always a conjunction of clauses, but may contain arbitrarily complex conjunctions or disjunctions nested within each clause. Instead of writing rules whose premise would be a disjunction of clauses, we write a separate rule for each clause. The action part indicates one or more conclusions which can be drawn if the premises are satisfied; hence the rules are (currently) purely inferential in character.

PREMISE   ($AND(SAME CNTXT INFECT PRIMARY-BACTEREMIA)
                            (MEMBF CNTXT SITE STERILESITES)
                            (SAME CNTXT PORTAL GI)

ACTION   (CONCLUDE CNTXT IDENT BACTEROIDES TALLY .7)

If   (1) the infection is primary-bacteremia, and
        (2) the site of the culture is one of the sterilesites, and
        (3) the suspected portal of entry of the organism is the gastrointestinal tract,
then there is suggestive evidence (.7) that the identity of the organism is bacteroides.

Figure 4. A rule from the knowledge base. $AND and $OR are the multivalued analogues of the standard Boolean AND and OR.

It is intended that each rule embody a single, modular chunk of knowledge, and state explicitly in the premise all necessary context. Since the rule uses a vocabulary of concepts common to the domain, it forms, by itself, a comprehensible statement of some piece of domain knowledge. As will become clear, this characteristic is useful in many ways.

Each rule is, as is evident, highly stylized, with the IF/THEN format and the specified set of available primitives. While the LISP form of each is executable code (and, in fact, the premise is simply EVALuated by LISP to test its truth, and the action EVALuated to make its conclusions), this tightly structured form makes possible the examination of the rules by other parts of the system. This in turn leads to some important capabilities, to be described below. For example, the internal form can be automatically translated into readable English, as shown in Figure 4.

Despite this strong stylization, we have not found the format restrictive. This is evidenced by the fact that of nearly 200 rules on a variety of topics, only 8 employ any significant variations. The limitations that do arise are discussed in Section 6.1.2.

3.3. Judgmental Knowledge

Since we want to deal with real-world domains in which reasoning is often judgmental and inexact, we require some mechanisms for being able to say that *"A suggests B,"* or *"C* and *D tend* to rule out *E."* The numbers used to indicate the strength of a rule (e.g., the .7 in Figure 4) have been termed Certainty Factors (CFs). The methods for combining CFs are embodied in a model of approximate implication. Note that while these are derived from and are related to probabilities, they are distinctly different (for a detailed review of the concept, see [30]). For the rule in Figure 4, then, the evidence is strongly indicative (.7 out of 1), but not absolutely certain. Evidence confirming an hypothesis is collected separately from that which disconfirms it, and the truth of the hypothesis at any time is the algebraic sum of the current evidence for and against it. This is an important aspect of the truth model, since it makes plausible the simultaneous existence of evidence in favor and against the same hypothesis. We believe this is an important characteristic of any model of inexact reasoning.

Facts about the world are represented as 4-tuples, with an associative triple and its current CF (Figure 5). Positive CF's indicate a predominance of evidence confirming an hypothesis, negative CFs indicate a predominance of disconfirming evidence.

Note that the truth model permits the coexistence of several plausible values for a single clinical parameter, if they are suggested by the evidence. Thus, for example, after attempting to deduce the identity of an organism, the system may have concluded (correctly) that there is evidence that the identity is E. coli and evidence that it is Klebsiella, despite the fact that they are mutually exclusive possibilities.

As a result of the program's medical origins, we also refer to the attribute part of the triple as 'clinical parameter,' and use the two terms interchangeably here. The object part (e.g., CULTURE-1, ORGANISM-2) is referred to as a context. This term was chosen to emphasize their dual role as both part of the associative triple and as a mechanism for establishing scope of variable bindings. As explained below, the contexts are organized during a consultation into a tree structure whose function is similar to those found in 'alternate world' mechanisms of languages like QA4.

3.4. Control Structure

The rules are invoked in a backward unwinding scheme that produces a depth-first search of an AND/OR goal tree (and hence is similar in some respects to PLANNER'S consequent theorems): given a goal to establish, we retrieve the (precomputed) list of all rules whose conclusions bear on the goal. The premise of each is evaluated, with each predicate function returning a number between −1 and 1. $AND (the multivalued analogue of the Boolean AND) performs a minimization operation, and $OR (sim-

---

(SITE CULTURE-1 BLOOD 1.0)
(IDENT ORGANISM-2 KLEBSIELLA .25)
(IDENT ORGANISM-2 E.COLI .73)
(SENSITIVS ORGANISM-1 PENICILLIN − 1.0)

Figure 5. Samples of information in the patient data base during a consultation.

ilar) does a maximization.[3] For rules whose premise evaluates successfully (i.e. greater than .2, an empirical threshold), the action part is evaluated, and the conclusion made with a certainty which is

$$\langle \text{premise-value} \rangle * \langle \text{certainty factor} \rangle$$

Those which evaluate unsuccessfully are bypassed, while a clause whose
truth cannot be determined from current information causes a new
subgoal to be set up, and the process recurs. Note that 'evaluating' here
means simply invoking the LISP EVAL function—there is no additional
rule interpreter necessary, since $AND, $OR, and the predicate functions
are all implemented as LISP functions.

3.4.1. Variations from the Standard Depth-First Search. Unlike PLANNER,
however, the subgoal which is set up is a generalized form of the original
goal. If, for example, the unknown clause is 'the identity of the organism
is E. coli', the subgoal which is set up is 'determine the identity of the
organism.' The new subgoal is therefore always of the form 'determine
the value of ⟨attribute⟩' rather than 'determine whether the ⟨attribute⟩ is
equal to ⟨value⟩'. By setting up the generalized goal of collecting all evidence about a clinical parameter, the program effectively exhausts each
subject as it is encountered, and thus tends to group together all questions
about a given topic. This results in a system which displays a much more
focussed, methodical approach to the task, which is a distinct advantage
where human engineering considerations are important. The cost is the
effort of deducing or collecting information which is not strictly necessary. However, since this occurs rarely—only when the ⟨attribute⟩ can
be deduced with certainty to be the ⟨value⟩ named in the original goal—
we have not found this to be a problem in practice.

A second deviation from the standard rule unwinding approach is that
every rule relevant to a goal is used. The premise of each rule is evaluated, and if successful, its conclusion is invoked. This continues until all
relevant rules have all been used, or one of them has given the result with
certainty. This use of all rules is in part an aspect of the model of judgmental reasoning and the approximate implication character of rules—
unless a result is obtained with certainty, we should be careful to collect
all positive and negative evidence. It is also appropriate to the system's
current domain of application, clinical medicine, where a conservative

[3]Note that, unlike standard probability theory, $AND does not involve any multiplication
over its arguments. Since CFs are not probabilities, there is no a priori reason why a product
should be a reasonable number. There is, moreover, a long-standing convention in work
with multi-valued logics which interprets AND as *min* and OR as *max* [20]. It is based
primarily on intuitive grounds: if a conclusion requires all of its antecedents to be true, then
it is a relatively conservative strategy to use the smallest of the antecedent values as the
value of the premise. Similarly, if any one of the antecedent clauses justifies the conclusion,
we are safe in taking the maximum value.

strategy of considering all possibilities and weighing all the evidence is preferred.

If, after trying all relevant rules (referred to as 'tracing' the subgoal), the total weight of the evidence about a hypothesis falls between $-.2$ and $.2$ (again, empirically determined), the answer is regarded as still unknown. This may happen if no rule were applicable, the applicable rules were too weak, the effects of several rules offset each other, or if there were no rules for this subgoal at all. In any of these cases, when the system is unable to deduce the answer, it asks the user for the value (using a phrase which is stored along with the attribute itself). Since the legal values for each attribute are also stored with it, the validity (or spelling) of the user's response is easily checked. (This also makes possible a display of acceptable answers in response to a '?' answer from the user.)

The strategy of always attempting to deduce the value of a subgoal, and asking only when that fails, would insure the minimum number of questions. It would also mean, however, that work might be expended searching for a subgoal, arriving perhaps at a less than definite answer, when the user already knew the answer with certainty. In response to this, some of the attributes have been labelled as LABDATA, indicating that they represent quantities which are often available as quantitative results of laboratory tests. In this case the deduce-then-ask procedure is reversed, and the system will attempt to deduce the answer only if the user cannot supply it. Given a desire to minimize both tree search and the number of questions asked, there is no guaranteed optimal solution to the problem of deciding when to ask for information, and when to try to deduce it. But the LABDATA—clinical data distinction used here has performed quite well, and seems to embody a very appropriate criterion.

Three other recent additions to the tree search procedure have helped improve performance. First, before the entire list of rules for a subgoal is retrieved, the system attempts to find a sequence of rules which would establish the goal with certainty, based only on what is currently known. Since this is a search for a sequence of rules with $CF = 1$, we have termed the result a *unity path*. Besides efficiency considerations, this process offers the advantage of allowing the system to make 'common sense' deductions with a minimum of effort (rules with $CF = 1$ are largely definitional). Since it also helps minimize the number of questions, this check is performed even before asking about LABDATA type attributes as well. Because there are few such rules in the system, the search is typically very brief.

Second, a straightforward bookkeeping mechanism notes the rules that have failed previously, and avoids ever trying to reevaluate any of them. (Recall that a rule may have more than one conclusion, may conclude about more than a single attribute, and hence may get retrieved more than once.)

Finally, we have implemented a partial evaluation of rule premises.

Since many attributes are found in several rules, the value of one clause (perhaps the last) in a premise may already have been established, even while the rest are still unknown. If this clause alone would make the premise false, there is clearly no reason to do all the search necessary to try to establish the others. Each premise is thus 'previewed' by evaluating it on the basis of currently available information. This produces a Boolean combination of TRUEs, FALSEs, and UNKNOWNs, and straight-forward simplification (e.g. $F \wedge U \equiv F$) indicates whether the rule is guaranteed to fail.

3.4.2. Templates. The partial evaluation is implemented in a way which suggests the utility of stylized coding in the rules. It also forms an example of what was alluded to earlier, where it was noted that the rules may be examined by various elements of the system, as well as executed. We require a way to tell if any clause in the premise is known to be false. We cannot simply EVAL each individually, since a subgoal which had never been traced before would send the system off on its recursive search.

However, if we can establish which attribute is referenced by the clause, it is possible to determine (by reference to internal flags) whether it has been traced previously. If so, the clause can be EVALed to obtain the value. A template (Figure 6) associated with each predicate function makes this possible. The template indicates the generic type and order of arguments to the predicate function, much like a simplified procedure declaration. It is not itself a piece of code, but is simply a list structure of the sort shown above, and indicates the appearance of an interpreted call to the predicate function. Since rules are kept in interpreted form (as shown in Figure 4), the template can be used as a guide to dissect a rule. This is done by retrieving the template for the predicate function found in each clause, and then using that as a guide to examining the clause. In the case of the function SAME, for instance, the template indicates that the clinical parameter (PARM) is the third element of the list structure which comprises the function call. The previewing mechanism uses the templates to extract the attribute from the clause in question, and can then determine whether or not it has been traced.

There are two points of interest here—first, part of the system is 'read-

Function	Template	Sample function call
SAME	(SAME CNTXT PARM VALUE)	(SAME CNTXT SITE BLOOD)

Figure 6. PARM is shorthand for clinical parameter (attribute); VALUE is the corresponding value; CNTXT is a free variable which references the context in which the rule is invoked.

ing' the code (the rules) being executed by another part; and second, this reading is guided by the information carried in components of the rules themselves. The ability to 'read' the code could have been accomplished by requiring all predicate functions to use the same format, but this is obviously awkward. By allowing each function to describe the format of its own calls, we permit code which is stylized without being constrained to a single form, and hence is flexible and much easier to use. We require only that each form be expressible in a template built from the current set of template primitives (e.g., PARM, VALUE, etc.). This approach also insures that the capability will persist in the face of future additions to the system. The result is one example of the general idea of giving the system access to, and an "understanding" of its own representations. This idea has been used and discussed extensively in [7].

We have also implemented antecedent-style rules. These are rules which are invoked if a conclusion is made which matches their premise condition. They are currently limited to common-sense deductions (i.e. CF = 1), and exist primarily to improve system efficiency. Thus, for example, if the user responds to the question of organism identity with an answer of which he is certain, there is an antecedent rule which will deduce the organism gramstain and morphology. This saves the trouble of deducing these answers later via the subgoal mechanism described above.

3.5. Meta-Rules

With the system's current collection of 200 rules, exhaustive invocation of rules would be quite feasible, since the maximum number of rules for a single subgoal is about 30. We are aware, however, of the problems that may occur if and when the collection grows substantially larger. It was partly in response to this that we developed an alternative to exhaustive invocation by implementing the concept of *meta-rules*. These are strategy rules which suggest the best approach to a given subgoal. They have the same format as the clinical rules (Figure 7), but can indicate that certain clinical rules should be tried first, last, before others, or not at all. Thus before processing the entire list of rules applicable to any subgoal, the meta-rules for that subgoal are evaluated. They may rearrange or shorten the list, effectively ordering the search or pruning the tree. By making them specific to a given subgoal, we can specify precise heuristics without imposing any extra overhead in the tracing of other subgoals.

Note, however, that there is no reason to stop at one level of meta-rules. We can generalize this process so that, before invoking any list of rules, we check for the existence of rules of the next higher order to use in pruning or rearranging the first list. Thus, while meta-rules are strategies for selecting clinical rules, second order meta-rules would contain information about which strategy to try, third order rules would suggest criteria for deciding how to choose a strategy, etc. These higher order rules represent a search by the system through "strategy space", and appear to be powerful constraints on the search process at lower levels. (We have not yet encountered higher order meta-rules in practice, but neither have we actively sought them.)

Note also that since the system's rule unwinding may be viewed as tree search, we have the appearance of a search through a tree with the interesting property that each branch point contains information on the best path to take next. Since the meta-rules can be judgmental, there exists the capability of writing numerous, perhaps conflicting heuristics, and having their combined judgment suggest the best path. Finally, since meta-rules refer to the clinical rules by their content rather than by name, the method automatically adjusts to the addition or deletion of clinical rules, as well as modifications to any of them.

The capability of meta-rules to order or prune the search tree has proved to be useful in dealing with another variety of knowledge as well. For the sake of human engineering, for example, it makes good sense to ask the user first about the positive cultures (those showing bacterial growth), before asking about negative cultures. Formerly, this design choice was embedded in the ordering of a list buried in the system code. Yet it can be stated quite easily and explicitly in a meta-rule, yielding the significant advantages of making it both readily explainable and modifiable. Meta-rules have thus proved capable of expressing a limited subset of the knowledge formerly embedded in the control structure code of the system.

Meta-rules may also be used to control antecedent rule invocation. Thus we can write strategies which control the depth and breadth of conclusions drawn by the system in response to a new piece of information.

```
PREMISE:     ($AND (MEMBF SITE CNTXT NONSTERILESITES)
                   (THEREARE OBJRULES(MENTIONS CNTXT PREMISE SAMEBUG)
ACTION:      (CONCLIST CNTXT UTILITY YES TALLY − 1.0)
```

If   (1) the site of the culture is one of the nonsterilesites, and
      (2) there are rules which mention in their premise a previous organism which may
          be the same as the current organism
Then it is definite (1.0) that each of them is not going to be useful.

Figure 7. A meta-rule. A previous infection which has been cured (temporarily) may reoccur. Thus one of the ways to deduce the identity of the current organism is by reference to previous infections. However, this method is not valid if the current infection was cultured from one of the non-sterile culture sites. Thus this meta-rule says, in effect, *if the current culture is from a non-sterile site, don't bother trying to deduce the current organisms identity from identities of previous organisms.*

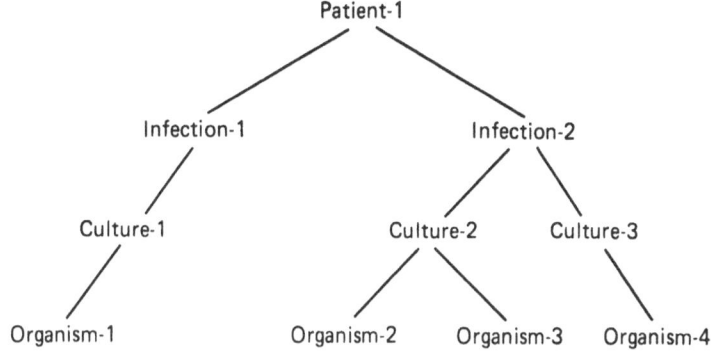

Figure 8. A sample of the contexts which may be sprouted during a consultation.

A detailed overview of all of these mechanisms is included in the Appendix, and indicates the way they function together to insure an efficient search for each subgoal.

The final aspect of the control structure is the tree of contexts (recall the dual meaning of the term, Section 3.3), constructed dynamically from a fixed hierarchy as the consultation proceeds (Figure 8). This serves several purposes. First, bindings of free variables in a rule are established by the context in which the rule is invoked, with the standard access to contexts which are its ancestors. Second, since this tree is intended to reflect the relationships of objects in the domain, it helps structure the consultation in ways familiar to the user. In the current domain, a patient has one or more infections, each of which may have one or more associated cultures, each of which in turn may have one or more organisms growing in it, and so on. Finally, we have found it useful to select one or more of the attributes of each context type and establish these as its MAIN-PROPS, or primary properties. Each time a new context of that type is sprouted, these MAINPROPS are automatically traced.[4] Since many of them are LABDATA-type attributes, the effect is to begin each new context with a set of standard questions appropriate to that context, which serve to 'set the stage' for subsequent questions. This has proved to be a very useful human engineering feature in a domain which has evolved a heavily stylized format for the presentation of information.

4. Relation to Other Work

We outline briefly in this section a few programs that relate to various aspects of our work. Some of these have provided the intellectual basis

---

[4]As a result of this, the control flow is actually slightly more complicated than a pure AND/ OR goal tree, and the flowchart in the appendix is correspondingly more complex.

from which the present system evolved, others have employed techniques which are similar, while still others have attempted to solve closely related problems. Space limitations preclude detailed comparisons, but we indicate some of the more important distinctions and similarities.

There have been a large number of attempts to aid medical decision making (see [27] for an extensive review). The basis for some programs has been simple algorithmic processes, often implemented as decision trees [23,37], or more complex control structures in systems tailored to specific disorders [2]. Many have based their diagnostic capabilities on variations of Bayes' theorem [10,36], or on techniques derived from utility theory of operations research [11]. Models of the patient or disease process have been used successfully in [32,25 and 17]. A few recent efforts have been based on some form of symbolic reasoning. In particular, the glaucoma diagnosis system described in [17] and the diagnosis system of [26] can also be viewed as rule-based.

Carbonell's work [5] represents an early attempt to make uncertain inferences in a domain of concepts that are strongly linked, much as MYCIN's are. Although the purpose of Carbonell's system was computer-aided instruction rather than consultation, much of our initial design was influenced by his semantic net model.

The basic production rule methodology has been applied in many different contexts, in attempts to solve a wide range of problems (see, for example, [6] for an overview). The most directly relevant of these is the DENDRAL system [3], which has achieved a high level of performance on the task of mass spectrum analysis. Much of the initial design of MYCIN was influenced by the experience gained in building and using the DENDRAL system, which in turn was based in part on [38].

There have been numerous attempts to create models of inexact reasoning. Among the more recent is [18], which reports on the implementation of a language to facilitate fuzzy reasoning. It deals with many of the same issues of reasoning under uncertainty that are detailed in [30].

The approach to natural language used in our system has been thus far quite elementary, primarily keyword-based. Some of the work reported in [4] suggested to us initially that this might be a sufficiently powerful approach for our purposes. This has proven generally true because the technical language of this domain contains relatively few ambiguous words.

The chess playing program of [41] employs a knowledge representation which is functionally quite close to ours. The knowledge base of that system consists of small sequences of code which recognize patterns of pieces, and then conclude (with a variable weighting factor) the value of obtaining that configuration. They report quite favorably on the ease of augmenting a knowledge base organized along these lines.

The natural language understanding system of [39] had some basic explanation capabilities similar to those described here, and could discuss its actions and plans.

As we have noted above, and will explore further below, part of our work has been involved in making it possible for the system to understand its own operation. Many of the explanation capabilities were designed and implemented with this in mind, and it has significantly influenced design of the knowledge acquisition system as well. These efforts are related in a general way to the long sequence of attempts to build program-understanding systems. Such efforts have been motivated by, among other things, the desire to prove correctness of programs (as in [35] or [22]), and as a basis for automatic programming (as in [13]). Most of these systems attempt to assign meaning to the code of some standard programming language like LISP, or ALGOL. Our attempts have been oriented toward supplying meaning for the terms used in MYCIN's production rules (such as SAME). The task of program-understanding is made easier by approaching it at this higher conceptual level, and the result is correspondingly less powerful. We cannot for instance prove that the implementation of SAME is correct. We can, however, employ the representation of meaning in other useful ways. It forms, for example, the basis for much of the knowledge acquisition program (see Section 6.3), and permits the explanation program to be precise in explaining the system's actions (see [7] for details). A similar sort of high level approach has been explored by Hewitt in his proposed INTENDER system [15].

Finally, similar efforts at computer-based consultants have recently been developed in different domains. The work detailed in [24] and [14] has explored the use of a consultation system similar to the one described here, as part of an integrated vision, manipulation, and problem solving system. Recent work on an intelligent terminal system [1] has been based in part on a formalism which grew out of early experience with the MYCIN system.

5. Fundamental Assumptions

We attempt here to examine some of the assumptions which are explicit and implicit in our use of production rules. This will help to suggest the range of application for these techniques, and indicate some of their strengths and limitations. Because such a listing is potentially open-ended, we include here the assumptions essential to the approach used in MYCIN, but which are not necessarily applicable to every interactive program.

There are several assumptions implicit in both the character of the rules and the ways in which they are used. First, it must be possible to write such judgmental rules. Not every domain will support this. It appears to require a field which has attained a certain level of formalization, which includes perhaps a generally recognized set of primitives and a minimal understanding of basic processes. It does not seem to extend to one which has achieved a thorough, highly formalized level, however. Assigning certainty factors to a rule should thus be a reasonable task whose results would be repeatable, but not a trivial one in which all CFs were 1.

Second, we require a domain in which there is a limited sort of interaction between conceptual primitives. Our experience has suggested that a rule with more than about six clauses in the premise becomes conceptually unwieldy. The number of factors interacting in a premise to trigger an action therefore has a practical (but no theoretical) upper limit. Also, the AND/OR goal tree mechanism requires that the clauses of a rule premise can be set up as non-conflicting subgoals for the purposes of establishing each of them (just as in robot problem solving; see [9] and the comment on side effects in [31]). Failure of this criterion causes results which depend on the order in which evidence is collected. We are thus making fundamental assumptions concerning two forms of interaction—we assume (a) that only a small number of factors (about 6) must be considered simultaneously to trigger an action; and (b) that the presence or absence of each of those factors can be established without adverse effect on the others.

Also, certain characteristics of the domain will influence the continued utility of this approach as the knowledge base of rules grows. Where there are a limited number of attributes for a given object, the growth in the number of rules in the knowledge base will not produce an exponential growth in search time for the consultation system. Thus as newly acquired rules begin to reference only established attributes, use of these rules in a consultation will not produce further branching, since the attributes mentioned in their premises will have already been traced. In addition, we assume that large numbers of antecedent rules will not be necessary, thus avoiding very long chains of 'forward' deductions.

There are essential assumptions as well in the use of this formalism as the basis for an interactive system. First, our explanation capabilities (reviewed below) rest on the assumption that display of either a rule or some segment of the control flow is a reasonable explanation of system behavior. Second, much of the approach to rule acquisition is predicated on the assumption that experts can be "debriefed," that is, they can recognize and then formalize chunks of their own knowledge and experience, and express them as rules. Third, the IF/THEN format of rules must be sufficiently simple, expressive, and intuitive that it can provide a useful language for expressing such formalizations. Finally, the system's mode of reasoning (a simple *modus ponens* chaining) must appear natural enough that a user can readily follow along. We offer below (Section 6) arguments that all these are plausible assumptions.

There is an important assumption, too, in the development of a system for use by two classes of users. Since the domain experts who educate the system so strongly influence its conceptual primitives, vocabulary, and knowledge base, we must be sure that the naive users who come for advice speak the same language.

The approach we describe does not, therefore, seem well suited to domains requiring a great deal of complex interaction between goals, or those for which it is difficult to compose sound judgmental rules. As a general indication of potentially useful applications, we have found that cognitive tasks are good candidates. In one such domain, antibiotic therapy selection, we have met with encouraging success.

6. Production Rules as a Knowledge Representation

In the introduction to this report we outlined three design goals for the system we are developing: usefulness (including competence), maintenance of an evolutionary knowledge base, and support of an interactive consultation. Our experience has suggested that production rules offer a knowledge representation that greatly facilitates the accomplishment of these goals. Such rules are straightforward enough to make feasible many interesting features beyond performance, yet powerful enough to supply significant problem solving capabilities. Among the features discussed below are the ability for explanation of system performance, and acquisition of new rules, as well as the general 'understanding' by the system of its own knowledge base. In each case we indicate the current performance levels of the system, and evaluate the role of production rules in helping to achieve this performance.

6.1. Competence

The competence of the system has been evaluated in two studies in the past few years. In mid-1974, a semi-formal study was undertaken, employing five infectious disease experts not associated with the project. They were asked to evaluate the system's performance on fifteen cases of bacteremia selected from current inpatients. We evaluated such parameters as the presence of extraneous questions, the absence of important ones, the system's ability to infer the identity of organisms, and its ability to select appropriate therapy. The principal problem discovered was an insufficient number of rules concerned with evaluating the severity of a patient's illness. Nevertheless, the experts approved of MYCIN's therapy recommendation in 72% of the evaluations. (There were also considerable differences of opinion regarding the best therapy as selected by the experts themselves.)

A more formal study is currently under way. Building on our experience gained in 1974, we designed a more extensive questionnaire and prepared detailed background information on a new set of fifteen patients. These were sent to five experts associated with a local hospital, and to five others across the country. This will allow us to evaluate performance, and in addition measure the extent to which the system's knowledge base reflects regional trends in patient care.

6.1.1. Advantages of Production Rules. Recent problem solving efforts in AI have made it clear that high performance of a system is often strongly correlated with the depth and breadth of the knowledge base. Hence, the task of accumulation and management of a large and evolving knowledge base soon poses problems which dominate those encountered in the initial phases of knowledge base construction. Our experience suggests that giving the system itself the ability to examine and manipulate its knowledge base provides some capabilities for confronting these problems. These are discussed in subsequent sections.

The selection of production rules as a knowledge representation is in part a response to this fact. One view of a production rule is as a modular segment of code [40], which is heavily stylized [38,3]. Each of MYCIN's rules is, as noted, a simple conditional statement: the premise is constrained to be a Boolean expression, the action contains one or more conclusions, and each is completely modular and independent of the others. Such *modular, stylized coding* is an important factor in building a system that is to achieve a high level of competence.

For example, any stylized code is easier to examine. This is used in several ways in the system. Initial integration of new rules into the knowledge base can be automated, since their premise and action parts can be systematically scanned, and the rules can then be added to the appropriate internal lists. In the question answering system, inquiries of the form 'Do you recommend clindamycin for bacteroides?' can be answered by retrieving rules whose premise and action contain the relevant items. Similarly, the detection of straightforward cases of contradiction and subsumption is made possible by the ability to examine rule contents. Stylized code also makes feasible the direct manipulation of individual rules, facilitating automatic correction of such undesirable interactions.

The benefits of modularized code are well understood. Expecially significant in this case are the ease of adding new rules and the relatively uncomplicated control structure which the modular rules permit. Since rules are retrieved because they are relevant to a specific goal (i.e., they mention that goal in their action part), the addition of a new rule requires only that it be added to the appropriate internal list according to the clinical parameters found in its action. A straightforward depth first search (the result of the backward chaining of rules) is made possible by the lack of interactions among rules.

These benefits are common to stylized code of any form. Stylization in the form of production rules in particular has proved to be a useful formalism for several reasons. In the domain of deductive problems, especially, it has proven to be a natural way of expressing knowledge. It also supplies a clear and convenient way of expressing modular chunks of knowledge, since all necessary context is stated explicitly in the premise. This in turn makes it easier to insure proper retrieval and use of each rule. Finally, in common with similar formalisms, one rule never directly calls another. This is a significant advantage in integrating a new rule into the system—it can simply be 'added to the pot' and no other rule need be changed to insure that it is called (compare this with the addition of a new procedure to a typical ALGOL-type program).

6.1.2. Shortcomings of Production Rules. Stylization and modularity also result in certain shortcomings, however. It is, of course, somewhat harder to express a given piece of knowledge if it must be put into a predetermined format. The intent of a few of the rules in our system is thus less than obvious to the naive user even when translated into English. The requirement of modularity (along with the uniformity of the knowledge base), means all necessary contextual information must be stated explicitly in the premise, and this at times leads to rules which have awkwardly long and complicated premises.

Another shortcoming in the formalism arises in part from the backward chaining control structure. It is not always easy to map a sequence of desired actions or tests into a set of production rules whose goal-directed invocation will provide that sequence. Thus, while the system's performance is reassuringly similar to some human reasoning behavior, the creation of appropriate rules which result in such behavior is at times nontrivial. This may in fact be due more to programming experience oriented primarily toward ALGOL-like languages, rather than any essential characteristic of production rules. After some experience with the system we have improved our skill at 'thinking backward.'

A final shortcoming arises from constraining rule premises to contain "pure" predicates.[5] This forces a pure problem reduction mode in the use of rules: each clause of a premise is set up as an independent goal, and execution of the action should be dependent solely on the success or failure of premise evaluation, without referencing its precise value. It is at times, however, extremely convenient to write what amounts to a 'for each' rule, as in 'for each organism such that . . . conclude . . . '. A few rules of this form are present in the system (including, for example, the meta-rule in Figure 7), and they are made to appear formally like the rest by allowing the premise to compute a value (the set of items that satisfy

---

[5]That is, a predicate that returns a value indicating only success or failure. Since we use a multi-valued logic, the predicate functions in rule premises return a number between 0 and 1. The alternative approach is to allow any non-NIL value to indicate success (e.g., the MEMBER function in LISP).

the premise), which is passed to the action clause via a global variable. While this has been relatively successful, the violation of the basic formalism results in other difficulties—in particular, in the explanation system, which produces somewhat murky explanations of such rules. We are working toward a cleaner solution of this problem.

6.2. Explanation

Augmentation or modification of any knowledge base is facilitated by the ability to discover what knowledge is currently in the system and how it is used. The system's acceptance (especially to a medical audience) will be strongly dependent upon the extent to which its performance is natural (i.e., human-like) and transparent. Lack of acceptance of some applications programs can be traced to their obscure reasoning mechanisms which leave the user forced to accept or reject advice without a chance to discover its basis. One of our original design criteria, then, was to give the system the ability to provide explanations of its behavior and knowledge. It soon became evident that an approach relying on some form of symbolic reasoning (rather than, for example, statistics) would make this feasible. This was one of the primary reasons behind the choice of the production rule representation, and has continued to influence the program's development.

Our initial efforts at explanation and question-answering were based on three capabilities: (i) display on demand during the consultation the rule currently being invoked, (ii) record rules which were invoked, and after the consultation, be able to associate specific rules with specific events (questions and conclusions) to explain why each of them happened, and (iii) search the knowledge base for a specific type of rule in answer to inquiries of the user. The first of these could be easily implemented via the single-word command format described below.

The latter two were intended for use after the consultation, and hence were provided with a simple natural language front end. Examples are shown in Figure 9 (additional examples can be found in [29]). Note that the capability for answering questions of type (ii) has been extended to include inquiries about actions the program *failed* to take (example [d], Figure 9). This is based on the ability of the explanation system to simulate the control structure of the consultation system, and can be extremely useful in deciphering the program's behavior. For questions of type (iii), ([e] in Figure 9) the search through the knowledge base is directed by a simple parsing of the question into a request for a set of rules, with constraints on premise and/or action contents. The retrieval of relevant rules is guided primarily by pre-established (but automatically generated) lists which indicate premise and action contents.

Some generalization of and extensions to the methodology of (i) and (ii) have been motivated by two shortcomings. Displaying the current

(a) •• **HOW DID YOU DECIDE THAT ORGANISM-1 MIGHT BE AN E.COLI?**

I used RULE021 to conclude that the identity of ORGANISM-1 is e.coli.
This gave a cumulative CF of (.47).

Then I used RULE084 to conclude that the identity of ORGANISM-1 is e.coli. This gave a cumulative CF of (.55).

Finally, I used RULE003 to conclude that the identity of ORGANISM-1 is e.coli. This gave a cumulative CF of (.74).

(b) •• **DID YOU USE RULE 163 TO FIND OUT ANYTHING ABOUT ORGANISM-1?**

RULE163 was tried in the context of ORGANISM-1, but it failed because it is not true that the patient has had a genito-urinary tract manipulative procedure (clause 3).

(c) •• **DID RULE 116 TELL YOU ANYTHING ABOUT INFECTION-1?**

No. RULE116 would be relevant to the primary-bacteremia when applied to ORGANISM-1.

But clause 2 (''the gram stain of the organism is not known'') was already known to be false for ORGANISM-1, so the rule was never tried.

(d) •• **WHY DIDN'T YOU CONSIDER STREPTOCOCCUS AS A POSSIBILITY FOR ORGANISM-1?**

The following rule could have been used to determine that the identity of ORGANISM-1 was streptococcus: RULE033

But clause 2 (''the morphology of the organism is coccus'') was already known to be false for ORGANISM-1, so the rule was never tried.

(e) •• **DO YOU RECOMMEND CLINDAMYCIN FOR BACTEROIDES?**

Yes.
RULE060
- - - - - - -

If the identity of the organism is bacteroides
Then I recommend therapy be chosen from among the following:
   CHLORAMPHENICOL, CLINDAMYCIN, RIFAMPIN, ERYTHROMYCIN, CARBENICIL-
   LIN, TETRACYCLINE, LINCOMYCIN

(f) •• **IS BLOOD A STERILE SITE?**

Yes, blood is one of those sites that are normally sterile.

(g) •• **WHAT ARE THE ORGANISMS ASSOCIATED WITH THE GASTRO-INTES-TINAL TRACT?**

The likely pathogens associated with the gastro-intestinal tract are:
ENTEROCOCCUS, CLOSTRIDIUM-GANGRENE, BACTEROIDES, KLEBISELLA, PSEU-
DOMONAS, E.COLI, ENTEROBACTER, PROTEUS.

Figure 9. Examples of natural language question answering capabilities. Questions [a]–[d] reference a specific consultation, while [e]–[g] are general inquiries answered from the system knowledge base.

rule is not particularly informative if the rule is essentially definitional and hence conceptually trivial. The problem here is the lack of a good gauge for the amount of information in a rule.

Recording individual rule invocations, questions and conclusions is useful, but as a record of individual events, it fails to capture the context and ongoing sequence. It is difficult therefore to explain any event with reference to anything but the specific information recorded with that event.

Two related techniques were developed to solve these problems. First, to provide a metric for the amount of information in a rule, we use (in a very rough analogy with information theory) $(-\log CF)$. Rules which are definitional $(CF = 1)$ have by this measure no information, while those which express less obvious implications have progressively more information. The measure is clearly imperfect, since first, CFs are not probabilities, and there is thus no formal justification that $-(\log CF)$ is a meaningful measure. Second, any sophisticated information content measure should factor in the state of the observer's knowledge, since the best explanations are those which are based on an understanding of what the observer fails to comprehend. Despite these shortcomings, however, this heuristic has proven to be quite useful.

To solve the second problem (explaining events in context) the process of recording individual rule invocations has been generalized: all the basic control functions of the system have been augmented to leave behind a history of each of their executions. This internal trace is then read by various parts of the explanation system to provide a complete, in-context explanation of any part of the system's behavior.

Because the consultation process is essentially one of search through an AND/OR goal tree, inquiries during the course of a consultation fall quite naturally into two types: WHY a question was asked, and HOW a conclusion was (or will be) reached. The first of these looks "up" the goal tree, in examining higher goals, while the second looks "down" in examining rules which may help achieve a goal. This part of the system's explanation capability can thus be viewed in general as a process of tree traversal. By combining this concept with the information content metric, we make possible explanations in varying levels of detail: the tree may be traversed in steps whose information content size is specified by the user, rather than simply stepping from goal to goal. At the start of the traversal process, "information distance" from the current goal to the top of the tree is normalized to 10. The argument to the WHY command (an integer between one and ten, assumed to be one if absent) is then taken to indicate some part of that distance. Thus, **WHY 3** indicates an explanation which encompasses approximately one-third of the total "conceptual" distance. Repeated WHY questions have the effect of stepping up the goal tree. Examples are shown below; additional examples are found in [7] and [29].

Similarly, **HOW** commands step "down" the tree. This can result in examining either branches which have already been traversed (as in the example below) or those which have yet to be tried (in which case the question becomes 'HOW will you determine . . . ').

The system's fundamental approach to explanation is thus to display some recap of its internal actions, a trace of its reasoning. The success of this technique is predicated on the claim that the system's basic approach to the problem is sufficiently intuitive that a summary of those actions is at least a reasonable basis from which to start. While it would be difficult to prove the claim in any formal sense, there are several factors which suggest its plausibility.

First, we are dealing with a domain in which deduction, and deduction in the face of uncertainty, is a primary task. The use of production rules in an IF/THEN format seems therefore to be a natural way of expressing things about the domain, and the display of such rules should be comprehensible. Second, the use of such rules in a backward chaining mode is, we claim, a reasonably intuitive scheme. *Modus ponens* is a well-understood and widely (if not explicitly) used mode of inference. Thus, the general form of the representation and the way it is employed should not be unfamiliar to the average user. More specifically, however, consider the source of the rules. They have been given to us by human experts who were attempting to formalize their own knowledge of the domain. As such, they embody accepted patterns of human reasoning, implying that they should be relatively easy to understand, especially for those familiar with the domain. As such, they will also attack the problem at what has been judged an appropriate level of detail. That is, they will embody the right size of "chunks" of the problem to be comprehensible.

We are not, therefore, recapping the binary bit level operations of the machine instructions for an obscure piece of code. We claim instead to be working with primitives and a methodology whose (a) substance, (b) level of detail, and (c) mechanism are all well suited to the domain, and to human comprehension, precisely because they were provided by human experts. This approach seems to provide what may plausibly be an understandable explanation of system behavior.

This use of symbolic reasoning is one factor which makes the generation of explanations an easier task. For example, it makes the display of a backtrace of performance comprehensible (as, for example, in Figure 9). The basic control structure of the consultation system is a second factor. The simple depth-first-search of the AND/OR goal tree makes HOW, WHY, and the tree traversal approach natural (as in Figures 10 and 11). We believe several concepts in the current system are, however, fairly general purpose, and would be useful even in systems which did not share these advantages. Whatever control structure is employed, the maintenance of an internal trace will clearly be useful in subsequent explanations of system behavior. The use of some information metric will help to insure that those explanations are at an appropriate level of detail. Finally, the explanation generating routines require some ability to decipher the actions of the main system.

By way of contrast, we might try to imagine how a program based on a statistical approach might attempt to explain itself. Such systems can, for instance, display a disease which has been deduced and a list of relevant symptoms, with prior and posterior probabilities. No more informative detail is available, however. When the symptom list is long, it

---

In this series, how many of the cultures were positive for the e.coli (ORGANISM-1)?
{typical system prompt}
•WHY
{request from user to determine motivation for question}
[i.e. WHY is it important to determine the number of cultures which were positive for the organism?]
{system's rephrasing of the WHY question, and answer}
[1.0] This will help to determine whether there is significant disease associated with this occurrence of the organism.
It has already been established that
[1.1] the culture was taken from a sterile source
[1.2] the number of cultures in the series is greater than 2
Therefore, if
[1.3] the number of cultures which were positive for ORGANISM-1 is greater than 2
Then there is suggestive evidence (.5) that there is significant disease associated with this occurrence of the e.coli (ORGANISM-1).
[RULE049]
••WHY
{repeated WHY to determine relevance of previous explanation}
[i.e. WHY is it important to determine if there is significant disease associated with this occurrence of the organism?]

[2.0] This will help to determine a diagnosis for which the infection should be treated.
It has already been established that
[2.1] the infection is not one of: primary-bacteremia, meningitis
[2.2] for some current culture of the patient it is true that the site of the culture is blood

Therefore, if
[2.3] there is significant disease associated with this occurrence of the organism
Then
it is definite (1.0) that the diagnosis for which the infection should be treated is secondary-bacteremia
[RULE 103]

Figure 10. Examples of explanation capabilities. User input is in **boldface** and follows the double asterisk. Expansion of each WHY or HOW question (enclosed in brackets) is produced by the system, to be sure the user is aware of the system's interpretation of them.

••HOW 1.1
                    {*the 1.1 refers to the information in the line labelled* [1.1] *in Fig. 10*}
[i.e. HOW was it determined that the culture was taken from a sterile source?]

[3.1] RULE 001 was used to conclude that there is suggestive evidence (.7) that the
      culture was taken from a sterile source.
[3.3] RULE 022 was used to conclude that there is strongly suggestive evidence (.7)
      that the culture was taken from a sterile source.
••HOW 3.1
                                                {*another request from the user*}
[i.e. HOW was RULE001 used?]
It has already been established that
[4.1] the site of the culture is one of: those sites from which the sterility of cultures
      depend upon the method of collection
[4.2] the method of collection of the culture is one of: the recognized methods of col-
      lection associated with the site of the culture, and
[4.3] it is not known whether care was taken in collecting the culture
Therefore
there is strongly suggestive evidence (.8) that the culture was taken from a sterile
source
[RULE022]

Figure 11. Examples of explanation capabilities—a HOW question.

may not be clear how each of them (or some combination of them) contributed to the conclusion. It is more difficult to imagine what sort of
explanation could be provided if the program were interrupted with
interim queries while in the process of computing probabilities. The
problem, of course, is that statistical methods are not good models of the
actual reasoning process (as shown in psychological experiments of [8]
and [33]), nor were they designed to be. While they are operationally
effective when extensive data concerning disease incidence are available,
they are also for the most part, "shallow", one-step techniques which capture little of the ongoing process actually used by expert problem solvers
in the domain.[6]

We have found the presence of even the current basic explanation
capabilities to be extremely useful, as they have begun to pass the most
fundamental test: it has become easier to ask the system what it did than

---

[6]However, the reasoning process of human experts may not be the ideal model for *all* knowledge-based problem solving systems. In the presence of reliable statistical data, programs
using a decision theoretic approach are capable of performance surpassing those of their
human counterparts.

In domains like infectious disease therapy selection, however, which are characterized
by "judgmental knowledge", statistical approaches may not be viable. This appears to be
the case for many medical decision making areas. See [12] and [30] for further discussion
of this point.

to trace through the code by hand. The continued development and generalization of these capabilities is one focus of our present research.

6.3. Acquisition

Since the field of infectious disease therapy is both large and constantly changing, it was apparent from the outset that the program would have to deal with an evolving knowledge base. The domain size made writing a complete set of rules an impossible task, so the system was designed to facilitate an incremental approach to competence. New research in the domain produces new results and modifications of old principles, so that a broad scope of knowledge-base management capabilities was clearly necessary.

As suggested above, a fundamental assumption is that the expert teaching the system can be "debriefed," thus transferring his knowledge to the program. That is, presented with any conclusion he makes during a consultation, the expert must be able to state a rule indicating all relevant premises for that conclusion. The rule must, in and of itself, represent a valid chunk of clinical knowledge.

There are two reasons why this seems a plausible approach to knowledge acquisition. First, clinical medicine appears to be at the correct level of formalization. That is, while relatively little of the knowledge can be specified in precise algorithms (at a level comparable to, say, elementary physics) the judgmental knowledge that exists is often specifiable in reasonably firm heuristics. Second, on the model of a medical student's clinical training, we have emphasized the acquisition of new knowledge in the context of debugging (although the system is prepared to accept a new rule from the user at any time). We expect that some error on the system's part will become apparent during the consultation, perhaps through an incorrect organism identification or therapy selection. Tracking down this error by tracing back through the program's actions is a reasonably straightforward process which presents the expert with a methodical and complete review of the system's reasoning. He is obligated to either approve of each step or to correct it. This means that the expert is faced with a sharply focussed task of adding a chunk of knowledge to remedy a specific bug. This makes it far easier for him to formalize his knowledge than would be the case if he were asked, for example, "tell me about bacteremia."

This methodology has the interesting advantage that the context of the error (i.e., which conclusion was in error, what rules were used, what the facts of this case were, etc.) is of great help to the acquisition system in interpreting the expert's subsequent instructions for fixing the bug. The error type and context supply the system with a set of expectations about the form and content of the anticipated correction, and this greatly facilitates the acquisition process (details of this and much of the operation of the acquisition system are found in [7]).

The problem of educating the system can be usefully broken down into three phases: uncovering the bug, transferring to the system the knowledge necessary to correct the bug, and integrating the new (or revised) knowledge into the knowledge base. As suggested above, the explanation system is designed to facilitate the first task by making it easy to review all of the program's actions. Corrections are then specified by adding new rules (and perhaps new values, attributes, or contexts), or by modifying old ones. This process is carried out in a mixed initiative dialogue using a subset of standard English.

The system's understanding of the dialog is based on what may be viewed as a primitive form of 'model-directed' automatic programming. Given some natural language text describing one clause of a new rule's premise, the system scans the text to find keywords suggesting which predicate function(s) are the most appropriate translations of the predicate(s) used in the clause. The appropriate template for each such function is retrieved, and the 'parsing' of the remainder of the text is guided by the attempt to fill this in.

If one of the functions were SAME, the template would be as shown in Figure 6. CNTXT is known to be a literal which should be left as is, PARM signifies a clinical parameter (attribute), and VALUE denotes a corresponding value. Thus the phrase "the stain of the organism is negative" would be analyzed as follows: the word "stain" in the system dictionary has as part of its semantic indicators the information that it may be used in talking about the attribute *gramstain* of an organism. The word "negative" is known to be a valid value of *gramstain* (although it has other associations as well). Thus one possible (and in fact the correct) parse is

(SAME CNTXT GRAM GRAMNEG)

or "the gramstain of the organism is gramnegative."

Note that this is another example of the use of higher level primitives to do a form of program understanding. It is the semantics of PARM and VALUE which guide the parse after the template is retrieved, and the semantics of the *gramstain* concept which allow us to insure the consistency of each parse. Thus by treating such concepts as conceptual primitives, and providing semantics at this level, we make possible the capabilities shown, using relatively modest amounts of machinery.

Other, incorrect parses are of course possible, and are generated too. There are three factors, however, which keep the total number of parses within reasonable bounds. First, and perhaps most important, we are dealing with a very small amount of text. The user is prompted for each clause of the premise individually, and while he may type an arbitrary amount at each prompt, the typical response is less than a dozen words. Second, there is a relatively small degree of ambiguity in the semi-formal language of medicine. Therefore a keyword-based approach produces only a small number of possible interpretations for each word. Finally, insuring the consistency of any given parse (e.g., that VALUE is indeed a valid value for PARM) further restricts the total number generated. Typically, between 1 and 15 candidate parses result.

Ranking of possible interpretations of a clause depends on expectation and internal consistency. As noted above, the context of the original error supplies expectations about the form of the new rule, and this is used to help sort the resulting parses to choose the most likely.

As the last step in educating the system, we have to integrate the new knowledge into the rest of the knowledge base. We have only recently begun work on this problem, but we recognize two important, general problems. First, the rule set should be free of internal contradictions, subsumptions, or redundancies. The issue is complicated significantly by the judgmental nature of the rules. While some inconsistencies are immediately obvious (two rules identical except for differing certainty factors), indirect contradictions (resulting from chaining rules, for example) are more difficult to detect. Inexactness in the rules means that we can specify only an interval of consistent values for a certainty factor.

The second problem is coping with the secondary effects that the addition of new knowledge typically introduces. This arises primarily from the acquisition of a new value, clinical parameter or context. After requesting the information required to specify the new structure, it is often necessary to update several other information structures in the system, and these in turn may cause yet other updating to occur. For example, the creation of a new value for the site of a culture involves a long sequence of actions: the new site must be added to the internal list ALLSITES, it must then be classified as either sterile or non-sterile, and then added to the appropriate list; if non-sterile, the user has to supply the names of the organisms that are typically found there, and so forth. While some of this updating is apparent from the structures themselves, much of it is not. We are currently investigating methods for specifying such interactions, and a methodology of representation design that minimizes or simplifies the interactions to begin with.

The choice of a production rule representation does impose some limitations in the knowledge transfer task. Since rules are simple conditional statements, they can at times provide power insufficient to express some more complex concepts. In addition, while expressing a single fact is often convenient, expressing a larger concept via several rules is at times somewhat more difficult. As suggested above, mapping from a sequence of actions to a set of rules is not always easy. Goal-directed chaining is apparently not currently a common human approach to structuring larger chunks of knowledge.

Despite these drawbacks, we have found the production rule formalism a powerful one. It has helped to organize and build, in a relatively short period, a knowledge base which performs at an encouraging level of competence. The rules are, as noted, a reasonably intuitive way of expressing simple chunks of inferential knowledge, and one which requires no acquaintance with any programming language. While it may not be immediately obvious how to restate domain knowledge in production rule format, we have found that infectious disease experts soon acquired some proficiency in doing this with relatively little training. We have had experience working with five different experts over the past few years, and in all cases had little difficulty in introducing them to the use of rules. While this is a limited sample, it does suggest that the formalism is a convenient one for structuring knowledge for someone unfamiliar with programming.

The rules also appear capable of embodying appropriate-sized chunks of knowledge, and of expressing concepts that are significant statements. They remain, however, straightforward enough to be built of relatively simple compositions of conceptual primitives (the attributes, values, etc.). While any heavily stylized form of coding of course makes it easier to produce code, stylizing in the form of production rules in particular also provides a framework which is structurally simple enough to be translatable to simple English. This means that the experts can easily comprehend the program's explanation of what it knows, and equally easily specify knowledge to be added.

7. Conclusions

The MYCIN system has begun to approach its design goals of competence and high performance, flexibility in accommodating a large and changing knowledge base, and ability to explain its own reasoning. Successful applications of our control structure with rules applicable to other problem areas have been (a) fault diagnosis and repair recommendations for bugs in an automobile horn system [34], (b) a consultation system for industrial assembly problems [14], and (c) part of the basis for an intelligent terminal system [1].

A large factor in this work has been the production rule methodology. It has proved to be a powerful, yet flexible representation for encoding knowledge, and has contributed significantly to the capabilities of the system.

Appendix

```
Procedure FINDVALUEOF (item GOAL)
    begin item X; list L; rule R; premise_clause P;
    if (X ← UNITYPATH(GOAL)) then return (X);
    if LABDATA(GOAL) and DEFINITE_ANSWER(X ← ASKUSER(GOAL)) then return(X);
    L ← RULES_ABOUT(GOAL);
    L ← APPLY_METARULES(GOAL,L,0);
```

```
for R ∈ L do
  unless PREVIEW(R) = false do
    begin "evaluate-rule"
    for P ∈ PREMISES_OF(R) do
      begin "test-each-premise-clause"
      if not TRACED(ATTRIBUTE_IN(P)) then FINDVALUEOF(ATTRIBUTE_IN(P));
      if EVALUATION_OF(P) < .2 then next(R);
      end "test-each-premise-clause";
    CONCLUDE(CONCLUSION_IN(R));
    if VALUE_KNOWN_WITH_CERTAINTY(GOAL) then
      begin MARK_AS_TRACED(GOAL); return(VALUEOF(GOAL)); end;
    end "evaluate-rule";
  MARK_AS_TRACED(GOAL);
  if VALUEOF(GOAL) = unknown and NOT_ALREADY_ASKED(GOAL)
                            then return(ASKUSER(GOAL))
                            else return(VALUEOF(GOAL));
end;


Procedure APPLY_METARULES(item GOAL; list L; integer LEVEL);
  begin list M; rule Q;
  if (M ← METARULES_ABOUT(GOAL,LEVEL+1))
                                        then APPLY_METARULES(GOAL,M,LEVEL+1);
  for Q ∈ M do USE_METARULE_TO_ORDER_LIST(Q,L);
  return(L);
  end;


Procedure CONCLUDE(action_clause CONCLUSION);
  begin rule T; list I;
  UPDATE_VALUE_OF(ATTRIBUTE_IN(CONCLUSION),VALUE_IN(CONCLUSION));
  I ← ANTECEDENTRULES_ASSOCIATED_WITH(CONCLUSION);
  I ← APPLY_METARULES(ATTRIBUTE_IN(CONCLUSION),I,0);
  for T ∈ I do CONCLUDE(CONCLUSION_IN(T));
  end;
```

The control structure as it might be expressed in an ALGOL-like language. All predicates
and functions refer to processes discussed in Sections 3.2–3.5. Examples of nonstandard
'data types' are shown below.

'data type'	example
item	IDENTITY
rule	RULE050
premise clause	(SAME CNTXT PORTAL GI)
action clause	(CONCLUDE CNTXT IDENTITY BACTEROIDES TALLY .8)

Acknowledgments. The work reported here was funded in part by grants from the Bureau of Health Sciences Research and Evaluation Grant HSO1544 and NIH Grant GM 29662, by a grant from the Advanced Research Projects Agency under ARPA Contract DAHC15-73-C-8435, and the Medical Scientist Training Program under NIH Grant GM-81922.

References

[1] Anderson, R.H. and Gillogly, J.J. RAND intelligent terminals agent: design philosophy, RAND R-1809-ARPA, The RAND Corporation, Santa Monica, CA. (January, 1977).

[2] Bleich, H.L. The computer as a consultant, *N. Engl. J. Med.* **284** (1971) 141–147.

[3] Buchanan, B.G. and Lederberg, J. The heuristic *DENDRAL* program for explaining empirical data, *IFIP* (1971), 179–188.

[4] Colby, K.M., Parkinson, R.C. and Faught, B.L. Pattern matching rules for the recognition of natural language dialog expressions, A I Memo #234, (June 1974), Computer Science Department, Stanford University, Stanford, CA.

[5] S. Carbonell, J.R. Mixed initiative man-computer instructional dialogues, BBN Report 1971, (31 May 1970), Bolt, Beranek and Newman, Cambridge, MA.

[6] Davis, R. and King, J. An overview of production systems, in: Elcock and Michie (Eds.), *Machine Intelligence* **8**: *Machine Representations of Knowledge*, Wiley, NY, 1977.

[7] Davis, R. Use of Meta-level knowledge in the construction and maintenance of large knowledge bases, A I Memo #283 (July 1976) Computer Science Department, Stanford University.

[8] Edwards, R. Conservatism in human information processing, in: Kleinmuntz (Ed.), *Formal Representation of Human Judgment*, Wiley, 1968, pp. 17–52.

[9] Fahlman, S.E. A planning system for robot construction tasks, *Artificial Intelligence* **5** (1974) 1–50.

[10] Gorry, G.A. and Barnett, G.O. Experience with a model of sequential diagnosis, *Comput. Biomed. Res.* **1** (1968) 490–507.

[11] Gorry, G.A., Kassirer, J.P., Essig, A. and Schwartz, W.B., Decision analysis as the basis for computer-aided management of acute renal failure, *Am. J. Med.* **55** (1973) 473–484.

[12] Gorry, G.A. Computer-assisted clinical decision making, *Method. Inform. Med.* **12** (1973) 42–51.

[13] Green, C.C., Waldinger, R.J., Barstow, D.R., Elschlager, R., Lenat, D.B., McCune, B.P., Shaw, D.E. and Steinberg, L.I. Progress report on program-understanding systems, A I Memo #240, (August 1974), Computer Science Department, Stanford University, Stanford, CA.

[14] Hart, P.E. Progress on a computer-based consultant, AI Group Technical Note 99, (January 1975), Stanford Research Institute, Menlo Park, CA.

[15] Hewitt, C., Procedure semantics—models of procedures and the teaching of procedures, in: Rustin, R. (Ed.), *Nat. Lang. Process.*, Courant Computer Science Symposium, **8** (1971) 331–350.

[16] Kagan, B.M., Fannin, S.L. and Bardie, F. Spotlight on antimicrobial agents—1973, *J. Am. Med. Assoc.* **226** (1973) 306–310.

[17] Kulikowski, C.A., Weiss, S. and Saifr, A., Glaucoma Diagnosis and Therapy by Computer, *Proc. Annu. Meet. Ass. Res. Vision Ophthamol.* (May 1973).

[18] Le Faivre, R.A. Fuzzy problem solving, Tech. Rept. 37, (September 1974), University of Wisconsin, Madison.

[19] Lesser, V.R., Fennell, R.D., Erman, L.D. and Reddy, D.R., Organization of the HEARSAY-II speech understanding system, *IEEE Trans. Acoust. Speech Signal Process.* ASSP-23, (February 1975), 11–23.

[20] Lukasiewicz, J.A. Numerical interpretation of the theory of propositions, in: Borkowski (Ed.), *Jan Lukasiewicz: Selected Works,* North-Holland, Amsterdam, 1970.

[21] The MACSYMA reference manual, The MATHLAB Group, (September 1974), Mass. Inst. Tech., Cambridge, MA.

[22] Manna, Z., Correctness of programs, *J. Comput. Syst. Sci.* (1969).

[23] Meyer, A.V. and Weissman, W.K. Computer analysis of the clinical neurological exam, *Comput. Biomed. Res.* 3 (1973) 111–117.

[24] Nilsson, N.J. (Ed.), Artificial Intelligence—Research and Applications, A I Group Progress Report, (May 1975), Stanford Research Institute, Menlo Park, CA.

[25] Pauker, S.G., Gorry, G.A., Kassirer, J.P. and Schwartz, W.B. Towards the simulation of clinical cognition: taking the present illness by computer, *Am. J. Med.* **60** (1976) 981–996.

[26] Pople, H., Meyers, J. and Miller, R. DIALOG, a model of diagnostic logic for internal medicine, *Proc. Fourth Internl. Joint Conf. Artificial Intelligence* Tiblisi, U.S.S.R. (September 1975), 848–855.

[27] Shortliffe, E.H., MYCIN: A rule-based computer program for advising physicians regarding antimicrobial therapy selection, A I Memo #251, (October 1974). Computer Science Department, Stanford University. Also to appear as *Computer-Based Medical Consultations: MYCIN,* American Elsevier, New York, 1976.

[28] Shortliffe, E.H., Axline, S.G., Buchanan, B.G. and Cohen, S.N. Design considerations for a program to provide consultations in clinical therapeutics, Presented at San Diego Biomedical Symposium, (February 6–8, 1974).

[29] Shortliffe, E.H., Davis, R., Buchanan, B.G., Axline, S.G., Green, C.C. and Cohen, S.N., Computer-based consultations in clinical therapeutics—explanation and rule acquisition capabilities of the MYCIN system, *Comput. Biomed. Res.* **8** (1975) 303–320.

[30] Shortliffe, E.H. and Buchanan, B.G. A model of inexact reasoning in medicine, *Math. Biosci.* **23** (1975) 351–379.

[31] Siklossy, L. and Roach, J. Proving the impossible is impossible is possible, *Proc. Third Internl. Joint Conf. Artificial Intelligence,* Stanford University, (1973) 383–387.

[32] Silverman, H. A digitalis therapy advisor, MAC TR-143, (January 1975), Project MAC, Mass. Inst. Tech., Cambridge, MA.

[33] Tversky, A. and Kahneman, D. Judgment under uncertainty: heuristics and biases, *Science,* **185** (1974) 1129–1131.

[34] van Melle, W. Would you like advice on another horn, MYCIN project internal working paper, (December 1974) Stanford University, Stanford, California.

[35] Waldinger, R. and Levitt, K.N. Reasoning about programs, *Artificial Intelligence* **5** (1974) 235–316.

[36] Warner, H.R., Toronto, A.F. and Veasy, L.G. Experience with Bayes' theorem for computer diagnosis of congenital heart disease, *Ann. N.Y. Acad. Sci.* **115** (1964) 558–567.

[37] Warner, H.R., Olmstead, C.M. and Rutherford, B.D. HELP—a program for medical decision-making, *Comput. Biomed. Res.* **5** (1972) 65–74.
[38] Waterman, D.A., Generalization learning techniques for automating the learning of heuristics, *Artificial Intelligence* **1** (1970) 121–170.
[39] Winograd, T. Understanding natural language, *Cognitive Psychology* **3** (1972).
[40] Winograd, T. Frame representations and the procedural/declarative controversy, in: Bobrow and Collins (Eds.), *Representation and Understanding,* Academic Press, 1975.
[41] Zobrist, A.L., Carlson, F.R. An advice-taking chess computer, *Sci. Am.* **228** (1973) 92–105.

Problems in the Design of Knowledge Bases for Medical Consultation

Casimir A. Kulikowski

In developing computer based medical consultation systems the choice of representation for the knowledge base crucially determines the scope, power and flexibility of the reasoning procedures that can be used with it. This paper describes the development of the CASNET/Glaucoma consultation system and discusses some concepts of knowledge base structure that arose during this process.

1. Introduction

Computer consultation systems that incorporate the knowledge and experience of expert clinicians can be flexible and useful tools for providing advice on complex cases of disease. When linked to a data base of representative cases in a specialty they become powerful tools for clinical research while also serving as instruments for health care and medical teaching. Computer systems that can reason explicitly with realistically complex structures of hypotheses, while providing explanations of their reasoning, are a recent development within artificial intelligence in medicine [1,2,3,4]. The construction of knowledge bases for these systems presents difficult problems because of the heterogeneity and complexity of medical knowledge. Differences of judgment among experts present further complications.

In this paper we describe the development of a consultation system (CASNET/Glaucoma [1,5,6]) that has reached a high level of clinical proficiency in providing advice on the diagnosis and management of difficult glaucoma cases. The description is used to highlight some of the problems that can arise in building a knowledge-based system.

2. Knowledge Bases for Consultation: Structure and Function

2.1. Normative Models

The specific outcomes that are desired of a consultation prescribe the structure and function of the knowledge bases that are needed to produce them. If a simple diagnostic statement is the only required output of a consultation system, it will be sufficient to specify the judgmental norms or rules that interpret a patient's medical history, observable manifestations and tests in terms of the diagnostic hypotheses. A general strategy of logical reasoning must also be chosen to specify the manner in which the rules are to be applied, in conjunction and/or in sequence, to produce a final diagnosis. Such a purely normative model of decision making represents the simplest and most efficient encoding of information for strictly diagnostic purposes. It has also been the earliest and most widely used scheme, since it characterizes the logical decision rule [7], statistical [8,9] and pattern recognition approaches [10] to diagnosis.

If a treatment recommendation constitutes the desired output of a consultation system, additional judgmental rules for selecting treatments on the basis of diagnostic and other patient information must be included in the knowledge base. In analogy to the diagnostic situation, a general strategy must then be chosen to specify the manner in which the treatment selection rules are to be applied to a given patient to produce a specific recommendation. If the alternative treatments are few and the cost/benefit tradeoff of the different outcomes can be defined, methods of statistical decision (or utility) theory can be applied [11,12].

The lack of large, reliable, and generally accepted data bases in medicine precludes the use of these simple normative models for any reasoning that involves large numbers of hypotheses and findings. A further limitation of such schemes is that their explanations are limited to listings of the findings that contribute to each hypothesis, and the extent of their contribution to the likelihood of the hypotheses. In many cases the statistical or heuristic functions used in hypothesis evaluation or treatment selection are complex enough that they represent "black boxes" whose internal workings defy clinical interpretation. If the accuracy of reasoning with such a system is exceptionally high this can be tolerated, but in most realistic situations it takes a long "training" period (often of several years) for a consultation system to reach acceptable levels of clinical performance (e.g. at a similar or better level than the performance of a predefined group of physicians). During this training a clear understanding of the reasons for each mistake of the system becomes crucial if it is to be easily corrected without creating new and unexpected errors in other parts of the reasoning. Explicitness of the reasoning procedures and modularity of the knowledge bases are thus necessary features of a system that is to be easily updated. These features are also necessary if the system's advice is to be accepted by the usually skeptical and exacting clinical users.

Systems that use artificial intelligence approaches have all stressed the importance of explanatory capabilities in consultation. One of these systems (MYCIN) uses a purely normative knowledge base. The others (CASNET/Glaucoma, INTERNIST, PIP) include various types of explicit descriptions of diseases. The MYCIN system uses a production rule formalism [13] to encode its decision rules, and has the power to represent a rich structure of intermediate hypotheses or actions leading up to a final decision. It does so by categorizing the different types of rules, and the elements that compose the premises and consequents of each rule. A uniform reasoning strategy is used with all rules, which is a recursive goal directed procedure, similar to that used in consequent theorem proving [2].

2.2. Descriptive Models

Purely normative models can express the decision rules of expert consultants in a very natural fashion. In doing so however, they usually combine within a rule, components that are purely judgmental together with others that are supported by scientific theories. A truly expert consultant should be able to support his reasoning with the latest research results, and make distinctions between those aspects that are justified by coherent theories and those that are based on empirical evidence alone. To make such distinctions within a computer-based consultation system, we require a descriptive model for disease processes and the episodes of illness that can arise from them. Only then can the system justify its reasoning in terms of known pathophysiological mechanisms of disease. Weaker aspects of a model, justified on phenomenological grounds alone and gaps in the knowledge base that must be bridged by subjective rules of clinical judgment can also be made explicit.

What constitutes a descriptive model of diseases, illnesses, their mechanisms and treatments? Quantitative models exist for many physiological mechanisms, but their clinical interpretation is difficult. It is usually in terms of qualitative descriptions, often of a causal nature, that the clinician justifies his reasoning. With this in mind, a descriptive scheme for modeling diseases can be conceived, in its simplest form, as a structure that contains observations and concepts layered in increasing order of abstraction and complexity (Figure 1).

The lower or least structured plane contains observations: signs, symptoms and test results. Relationships at this level are purely statistical or are based on phenomenological constraints. The second plane contains the anatomical-physiological states which serve as summaries of pathophysiological conditions that can be linked together by cause-and-effect and other semantic relationships to describe the dynamic process of dis-

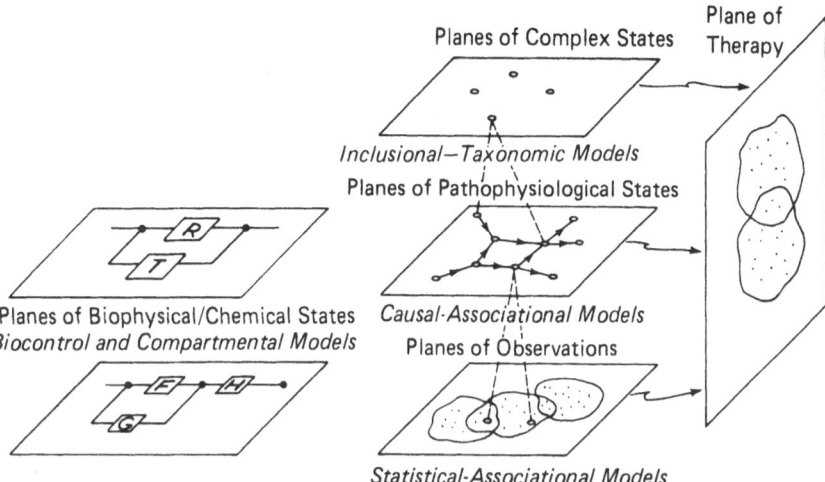

Figure 1. A representation of models of disease.

ease evolution. These states are often not directly observable or measurable entities, but must be inferred from the specific observations made at level one. Typical states, taken from the CASNET/Glaucoma model, would be "increased intraocular pressure," "cupping of the optic disk" and "visual field loss." The upper and most structured level contains complex states such as mechanisms, syndromes and diseases.

These can be linked by inclusional and other semantic relationships (such as: angle closure glaucoma is-a-subcategory-of glaucoma) and structured into a taxonomic network. Some of these complex states will be linked by causal relations, and therefore also enter into the causal-associational network. Others will be mechanisms of dysfunction that can be described by pathways through or subgraphs of the causal net. Thus, although only two levels of states are illustrated in the figure, they stand for a continuum of states defined at increasing levels of abstraction.

An alternative representation is illustrated by the planes to the left, containing free variables that enter into the mathematical models of biophysical and biochemical systems. The main problem in utilizing such models in consultative reasoning is that their scope is often too limited and their parameters difficult to evaluate outside of controlled laboratory experiments. Thus, links from these models to the clinically useful information at the level of the causal network are restricted to only subgraphs of the entire network graph.

A distinct plane contains the mechanisms of therapy. It can be visualized as being orthogonal to the others because it encompasses a qualitatively different category: that of actions. Once therapy is instituted it changes many of the relationships between the pathophysiological states, which are also reflected at the level of observations.

The strategies for diagnostic interpretation, prediction, and treatment selection can be designed to take advantage of the strongest of the various types of relationships encountered in the knowledge base. These, in turn, depend on the nature of the knowledge in the medical specialty that is being modeled. For diseases whose mechanisms are known, comparison with a causal sequence of states can help trace the development of an illness, assess its stage of progression, and predict future effects. This approach has been emphasized in the CASNET model. In situations where such detailed knowledge is lacking, greater reliance must be placed on the taxonomic structuring of the diseases and states. Because of the judgmental arbitrariness that characterizes all classification schemes, purely taxonomic models lack uniqueness. Nevertheless, taxonomic relations are essential for defining states of dysfunction at different levels of abstraction, and these are a necessary ingredient in the formulation of complex hypotheses.

3. Causal-Associational Network (CASNET) Models

A causal-associational network is a particular type of semantic network designed to describe disease mechanisms by causal pathways, relate these to an associational structure of external observations, and enable various classification structures to be imposed on the network for the purpose of defining diagnostic, prognostic, and treatment categories.

3.1. Causal Network of States

In our model of disease, the pathogenesis and mechanisms of a disease process are described in terms of cause and effect relationships between pathophysiological states. States are summary descriptions of events which are deviations from normality. Strict causality is not assumed— there may be multiple causes and effects, and in a given patient, a cause may be present without any of its effects occurring at the same time. Various effects can follow from a given cause, each produced with a different strength of causation. Many states may occur simultaneously in any disease process. A state thus defined may be viewed as a qualitative restriction on a state variable as used in control systems theory. It does not correspond to one of the mutually exclusive states that can be used to describe a probabilistic system. This definition was chosen to correspond to the basic entities physicians appear to use when they describe disease mechanisms. Disease processes may be characterized by pathways through the network. A complete pathway from a starting to a terminal node usually represents a complete disease process while partial pathways, from starting to nonterminal nodes, represent various degrees of evolution within the disease process. Progression along a causal pathway is usually associated with increasing seriousness of the disease.

When a set of cause and effect relationships between states is specified, the resulting structure is a network, or directed acyclic graph of states.

The mappings between nodes n_i of the causal net are of $n_i \xrightarrow{a_{ij}} n_j$ where a_{ij} is the strength of causation (interpreted in terms of its frequency of occurrence and n_i and n_j are states which are summarized by English language statements. This rule is interpreted as: state n_i causes state n_j, independent of other events, with frequency a_{ij}. Starting states are also assigned a frequency measure indicating a prior or starting frequency. The levels of causation are represented by numerical values, fractions between zero and one, which correspond to qualitative ranges such as: sometimes, often, usually, or always. States are summary statements. Many events and many complex relationships may be summarized by a single state. For example, "neutral tissue loss and cupping of the nerve head" is a summary of a much more complex situation. If a higher resolution description is desired, several different types of nerve loss and cupping could be specified. The resolution of states should be maintained at a level consistent with the objective of efficient decision-making. A state network can be thought of as a streamlined model of disease that unifies several important concepts and guides us in our goal of diagnosis. It is not meant to be a complete model of disease.

3.2. Rules for Associating States to Observations

Observations (tests)—the history, signs, symptoms, and laboratory tests—are the form in which information about a patient is presented. These clinical features, however, must be unified into some coherent framework for explanation and diagnosis. Observations about a patient are used to confirm or deny certain states in the network which describe the disease process. A single state may summarize many observations. These states can then be related by causal pathways that explain the mechanisms of disease in that patient.

The relationship between tests and states is non-causal; it is associational. For a given observation, confidence measures are used to indicate a degree of belief in the presence of specific states.

The rules for associating tests with states are represented as:

$$t_i \xrightarrow{Q_{ij}} n_j$$

where t_i is the i-th observation (or Boolean combination of observations), n_j is the j-th node, and Q_{ij} ($-1 \leq Q_{ij} \leq +1$) is the confidence in n_j given that t_i is observed to be true. Positive values of Q_{ij} correspond to an increased confidence in n_j and negative values correspond to a decreased confidence in n_j when t_i is observed. Associated with each observation are costs $C(t_i)$ that reflect the cost of obtaining the result t_i.

3.3. Rules for Associating Disease Categories to States

Diagnostic and prognostic categories of disease are defined in terms of ordered patterns of states, which we refer to as classification tables.

The tables contain rules of the form:

$$n_1 \wedge \quad \bar{n}_2 \qquad\qquad\qquad \rightarrow D_1$$
$$n_1 \wedge \quad n_2 \wedge \qquad\quad \bar{n}_3 \quad \rightarrow D_2$$

$$\cdot$$
$$\cdot$$
$$\cdot$$

$$n_1 \wedge \quad n_2 \wedge \ldots \quad \bar{n}_i \quad \rightarrow D_{i-1}$$
$$n_1 \wedge \quad n_2 \wedge \ldots \quad n_i \quad \rightarrow D_i$$

Each basic cause or starting state in the causal network, has pointers to the particular classification tables that contain diagnostic statements which describe its disease mechanism. Several starting states may refer to the same table, since several causal mechanisms may be included in the same diagnostic category.

Within a table, differing intensities of a disease process can be identified by differences in the magnitude or intensity of its characteristic states. In glaucoma, different intensities of pressure may be distinguished by defining states of moderately elevated pressure or extremely elevated pressure. These states, when found in classification tables, may then lead to different conclusions. The most effective use of the classification ordering, however, lies in making it correspond to the progression of states of a disease, characterized by following the states in a causal pathway.

3.4. Reasoning Strategies for CASNET Models

Diagnosis is implemented by first interpreting the patient's observations in terms of their underlying pathophysiological states. The relationships between observations and states are summarized by measures of confidence in the states for a given pattern of observations. A three-valued logic (confirmed, denied, undetermined) is used to summarize the truth value of each pathophysiological state taken as a hypothesis for the patient. Diagnoses are then triggered by various configurations of confirmed or denied states within the causal network. Prognoses result from extrapolating to possible future effects along confirmed causal pathways. Therapies are associated with the control of causative states and related patterns of significant observations. They are ranked by indices of severity (toxicity, side effects) and effectiveness, and are selected under a criterion of minimal severity within the class of effective therapies for each diagnostic or prognostic state.

When requested by the user, the CASNET/Glaucoma system can back up its reasoning by selected quotations and references from the current literature, including statistical summaries of results from research studies, and alternative opinions when a subject is currently under debate.

3.5. Development of the CASNET/Glaucoma Consultation System

The design of a consultation system can be broken down into two important tasks. These are the design and representation of models, and the design of general problem solving algorithms which use a suitably defined model for decision-making. In the general CASNET system, a user describes or modifies a model, but does not alter the reasoning procedures that select diagnostic interpretation and treatment plans. Two separate computer programs have been developed: the modeling program for designing application models [14], and the consultation program that uses models for reaching diagnoses and recommending therapies [6].

The current glaucoma consultation system has more than 100 states, 400 tests, 75 classification tables, and 200 diagnostic and treatment statements. Results must be interpreted for each eye, so that, in effect, twice the number of rules are involved in any ophthalmological model. There are also many special rules for binocular comparisons of states, tests, and diagnostic and treatment statements. A set of the program's conclusions for a sample case is given in Figure 2. This session illustrates the level of performance which the program has attained in reasoning about complex cases of glaucoma.

The consultation program has been designed for efficient performance. Human engineering aspects of program design have also been emphasized. The program has been developed primarily as a tool for research of medical decision-making by computer. However, our approach to program development involved the collaboration of a network of physicians with minimal prior experience in the use of computers. Their active participation in the project required careful attention to programming details which would allow our collaborators and other ophthalmologists to use the programs with little difficulty. This implies that only limited typing would be required, and quick response time, even for complex diagnostic interpretations would be essential.

Initially, we designed and built a prototype model which was demonstrated to a select audience of ophthalmologists. At this point, the program was far from being expert. However, rapid progress in the development of a decision-making system can be made by building a small simplified prototype and modifying and improving the prototype.

A very significant event in the development of the program has been the formation of ONET—the Ophthalmological Network. Using the SUMEX-AIM computer, this nationwide group of ophthalmological clinician-researchers have participated in the development of the program's knowledge base. They enter cases and suggest improvements. Their sug-

```
***********************
*DIAGNOSIS AND THERAPY*
***********************
```

*****VISIT 1:

* RIGHT EYE: *

(1) PRESENT DIAGNOSTIC STATUS:

PIGMENTARY GLAUCOMA. OPEN ANGLE GLAUCOMA. CHARACTERISTIC VISUAL FIELD LOSS WITH CORRESPONDING DISC CHANGES. EARLY FIELD LOSS.

(2) TREATMENT RECOMMENDATIONS:

PILOCARPINE 2% QID.

```
****RESEARCH STUDIES****
```

ALTERNATIVE INTERPRETATIONS OF PIGMENTARY GLAUCOMA:
    .SECONDARY GLAUCOMA
    .PRIMARY OPEN ANGLE GLAUCOMA
REFERENCES:
1. "WHEN PIGMENTARY GLAUCOMA WAS FIRST DESCRIBED IT WAS THOUGHT TO BE A FORM OF SECONDARY GLAUCOMA CAUSED BY PLUGGING OF THE TRABECULAR MESHWORK BY THE SAME PIGMENT THAT FORMED THE KRUKENBERG'S SPINDLES. HOWEVER, AN INCREASING NUMBER OF OBSERVERS NOW BELIEVE THAT IT IS A VARIANT OF PRIMARY OPEN-ANGLE GLAUCOMA. . . . (WILENSKY, PODOS—1975, TRANSACTIONS-NEW ORLEANS ACAD. OPHTH.)
2.   MORE RECENT EVIDENCE SUGGESTS THAT PIGMENTARY GLAUCOMA IS A SEPARATE ENTITY . . . (ZINK,PALMBERG, ET AL. A.J.O., SEPT. 1975)

                                    .

                                    .

*****VISIT 7:

* RIGHT EYE: *

(1) PRESENT DIAGNOSTIC STATUS:

PIGMENTARY GLAUCOMA. OPEN ANGLE GLAUCOMA. CHARACTERISTIC VISUAL FIELD LOSS WITH CORRESPONDING DISC CHANGES. ADVANCED FIELD LOSS. CURRENT MEDICATION HAS NOT CONTROLLED IOP IN THIS EYE. (AS INDICATED BY PROGRESSION OF CUPPING) (AS INDICATED BY VISUAL FIELD LOSS PROGRESSION)

(2) TREATMENT RECOMMENDATIONS:

FILTERING SURGERY IS INDICATED. AS AN ALTERNATIVE, PHOSPHOLINE MAY BE TRIED (BUT NOT USED 2 WEEKS BEFORE SURGERY).

Figure 2. Example of diagnosis and therapy for a case of pigmentary glaucoma, abstracted from a sequence of seven visits.

gestions have not been based on a comprehensive review of the logical rules contained in the program. Rather, we have concentrated on entering realistic cases and comparing the program's questioning sequence and conclusions with those of the experts.

Within a period of approximately a year and a half of ONET collaboration, the program achieved an expert level in the long term diagnosis and treatment of many types of glaucoma. The program's performance has been validated by our group of experts and by the system's participation in panel discussions of glaucoma cases at ophthalmological symposia. In November 1976 a scientific exhibit was given of the program at the annual meeting of the American Academy of Ophthalmologists and Otolaryngologists. Ophthalmologists were invited to present difficult cases to the computer. The program did well, with 95% of the ophthalmologists accepting the competency of its performance. Of these 77% described the program as performing at an expert or very competent level [6].

Acknowledgments. The author wishes to thank his colleague Dr. Sholom Weiss for numerous discussions on the topic of this paper. The work was supported in part by RR643 of Biotechnology Resources Program, DRR-NIH.

References

[1] Kulikowski, C. and Weiss S., "Computer-based Models of Glaucoma," Computers in Biomedicine Technical Report No. 3. Department of Computer Science, Rutgers University, 1971.

[2] Shortliffe, E.H., Axline, S.G., Buchanan, B.G., Merigan, T.C., and Cohen, S.N., An Artificial Intelligence Program to Advise Physicians Regarding Antimicrobial Therap. *Computers in Biomedical Research* 6, 544 (1973).

[3] Pople, H., Myers, J., Miller, R., Dialog: A Model of Diagnostic Logic for Internal Medicine, Proceedings of Fourth International Joint Conference on Artificial Intelligence, pp. 841–855 (1975).

[4] Pauker, S., Gorry, G., Kassirer, J. Schwartz, W., Towards the Simulation of Clinical Cognition—Taking a Present Illness by Computer, *American Journal of Medicine,* Vol. 60, pp. 981–996, 1976.

[5] Weiss, S., A System for Model-based Computer-aided Diagnosis and Therapy. Ph.D Dissertation, Rutgers University, 1974.

[6] Weiss, S., Kulikowski, C., Glaucoma Consultation by Computer, *Computers in Biology and Medicine,* in press.

[7] Ledley, R.S., and Lusted, L.B., Reasoning Foundation of Medical Diagnosis: Symbolic Logic, Probability and Value Theory and Our Understanding of How Physicians Reason, *Science,* 130, 1959.

[8] Warner, H.R., Toronto, A.F., and Veasy, L.G., Experience with Bayes' Theorem for Computer Diagnosis of Congenital Heart Disease. *Ann. N.Y. Acad. Sciences.* 115, 558, (1964).

[9] Gorry, G.A. and Barnett, G.O., Experience with a Model of Sequential Diagnosis. *Computers and Biomedical Research* 1, 490, (1968).

[10] Patrick, E.A., Stelmack, F.P., and Shen, L.Y.L. Review of Pattern Recognition in Medical Diagnosis and Consulting Relative to a New System Model. *IEEE Transactions on Systems, Man, and Cybernetics,* 4, 1974, 1–16.

[11] Ginsberg, A.S. and Offensend, F.L., An Application of Decision Theory to a Medical Diagnosis-Treatment Problem, *IEEE Transactions* Vol. SSC-4, No. 3, pp. 355–359, 1968.

[12] Schwartz, W.B., Gorry, G.A., Kassirer, J.P. and Essig, A., Decision Analysis and Clinical Judgment, *American Journal of Medicine,* pp. 459–472, 1973.

[13] Davis, R. and King, J., An Overview of Production Systems, Stanford Univ. Computer Science Department, Report No. STAN-CS-75-524, Stanford Artificial Intelligence Laboratory Memo AIM-271, 1975.

[14] Kulikowski, C., Weiss, S., An Interactive Facility for the Inferential Modeling of Disease, in Proceedings of 1973 Princeton Conference on Information Sciences and Systems, p. 524.

A Production Rule System for Neurological Localization

James A. Reggia

A rule-based program for localization of damage to the central nervous system was developed to evaluate MYCIN-like production system methodology. The program uses the results of the neurological examination of unconscious patients to categorize them in a manner familiar to clinicians. A collection of rules was found to be a poor representation for neurological localization knowledge because such information is conceptually organized in a frame-like fashion and is very context-dependent. Rule understandability was improved through the use of "macropredicates" and by the development of a natural inference hierarchy. The role of production systems as a model of human cognition is discussed.

Introduction

The creation of "expert systems," both in medicine and in various non-medical areas, is an active topic of research in the field of artificial intelligence. In medical applications, the basic idea behind these programs is to capture diagnostic and therapeutic decision-making knowledge in a computer program. The expectation is that such automated expertise, together with the computer's powerful logic and memory resources, can then be applied to many useful tasks. Examples of such tasks include: assisting physicians with the complexities of medical diagnosis, improving the quality of clinical education, and improving information retrieval from medical databases.

A major problem faced by those designing expert AI systems in medicine is that of *knowledge representation,* i.e., how to represent clinical diagnostic knowledge in a form suitable for computer processing. One solution, the representation of medical knowledge as a collection of *productions* or *rules,* has generated increasing interest over the last several years. Diagnostic programs using this methodology have important advantages over others in terms of modularity, comprehensibility to physicians, and educational utility. They are felt by some to be a reasonably accurate model of medical decision-making. The need to evaluate such rule-based systems and their associated inference methods further, especially in terms of their generality as models of the diagnostic process, has recently been emphasized (Shortliffe, 1976) and served as the impetus for the study reported here.

Background

Production Systems

Production systems (PS's) constitute a programming methodology that has been used in many ways, most notably for modeling symbol-processing aspects of cognition (Newell and Simon, 1972) and implementing expert, applications-oriented programs. This latter usage is what concerns us here. It is useful to think of such expert PS's as having three main components (Davis and King, 1976): a database, a set of rules, and a rule interpreter.

The *database* in a PS represents the state of the problem being solved, and thus in medical diagnostic programs, forms a record of the patient's signs and symptoms, as well as inferences made about the patient. Because of the uncertainties involved in diagnosis, facts about the patient are often stored along with some measure of their probability of being true.

The *rules* (or *productions*) in a PS represent the expert knowledge about problem solving in the domain to which the program is to be applied. They commonly take the form

$$\text{antecedents} \xrightarrow{x} \text{consequents}$$

meaning that if certain *antecedents* are true according to the database, then it logically follows that the *consequents* are also true with a "certainty" of x. By expressing knowledge in such discrete rules, one has the advantages of modularity (easy changeability and extensibility) and a potentially powerful explanation capability (the program can explain its decisions by supplying the relevant rules it used to make them).

Finally, the *rule interpreter* in a PS selects which rules to use at any particular time. This can be *antecedent-driven* where the occurrence of one or more of the antecedents in the database "triggers" the application of the rule to infer its consequents, or *consequent-driven* where the interpreter, in attempting to establish a certain fact, selects a rule with that

	Domain	Name (Source)
Nonmedical:	Chemical Spectra	DENDRAL (Buchanan, et al, 1970)
	Automatic Programming	PECOS (Barstow, 1977b)
	Geology	PROSPECTOR (Duda, et al, 1977)
	Molecular Genetics	MOLGEN (Martin, et al, 1977)
	X-ray Crystallography	SU/P (Nii, et al, 1978)
Medical:	Infections	MYCIN (Shortliffe, 1976)
	Glaucoma	CASNET/GLAUCOMA (Weiss, et al, 1978)
	Chorioretinal Disease	MEDICO (Walser, et al, 1977)

Figure 1. Example expert rule-based systems.

fact as a consequent and then tries to verify it by confirming that the antecedents are in the database. Such an associative approach to control structure is generally referred to as *pattern-directed* inferencing (Hayes-Roth, et al., 1978).

Figure 1 lists several examples of expert PS's to illustrate the variety of tasks to which the methodology is being applied. A more extensive discussion of the PS method can be found in the references given there (and in: Davis and King, 1976; Waterman and Hayes-Roth, 1978; Davis, et al., 1977).

Neurological Localization and its Automation

Only a brief, oversimplified outline of neurological localization and its application in establishing the cause of coma in a patient is given here for non-medical readers. Again, more extensive discussion is available in the cited references (Plum and Posner, 1972; Vinken and Bruyn, 1967; Haymaker, 1969).

Neurological localization is the determination of the loci of nervous system damage in a patient with neurological problems. It is achieved in two steps: first, a careful neurological examination is conducted where data is collected (testing of mental status, reflexes, muscle strength, etc.); and, second, this data is analyzed in order to determine the site or sites of brain damage most likely to explain the examination findings. Neurological localization is a necessary prerequisite for making an adequate diagnosis in a patient with a disease of the nervous system.

In the case of a patient in a coma of unknown etiology, the neurological examination and subsequent lesion localization is of critical importance. This is because the different disease processes that produce unconsciousness can involve different regions of the nervous system. Accurate

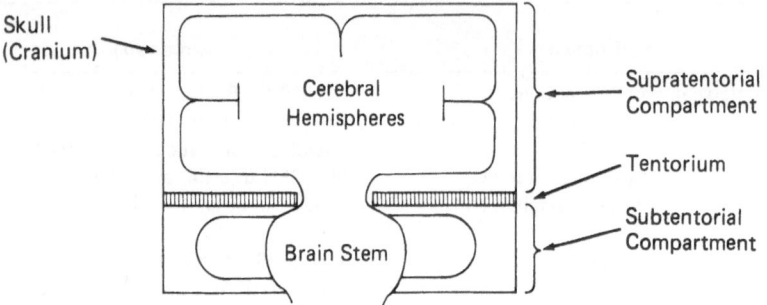

Figure 2. Schematic cross-sectional diagram of head to show how the tentorium divides the intracranial space into two compartments.

localization thus helps with the selection of a likely diagnosis and appropriate treatment.

For diagnostic purposes, it is very useful to classify unconscious patients (this is a simplification) into the following four categories (refer to Figure 2):

1. *Subtentorial lesion:* focal damage to the brainstem (the brainstem lies below a structure called the tentorium);
2. *Supratentorial lesion:* focal damage to a cerebral hemisphere (the hemispheres lie above the tentorium);
3. *Transtentorial herniation:* the patient has a supratentorial mass lesion (i.e., space-occupying) that takes up so much room inside the skull that the brain is displaced downward causing it to herniate (i.e., pass through) the tentorium (hence "transtentorial");
4. *Nonfocal:* diffuse impairment of the nervous system (e.g., sedative overdose).

While a great deal of research has been invested in studying automated diagnosis in the past, very little attention has been given to computer-assisted localization of neurological lesions. Bayes' theorem has been used to predict the site and histological type of intracerebral lesions using EEG, isotope scan, arteriography, and other clinical data (Stewart and Cala, 1975; Virnich, et al., 1975). Another study attempted to localize lesions in the brainstem by computer analysis of the neurological examination (Meyer and Weissman, 1973a and 1973b). This latter program considered a series of examination results and used a three-dimensional array (in which the brainstem was conceptually imbedded) to calculate the probability that each matrix element was involved by a lesion. The program's output was thus a display of matrix probabilities and its methodology suffered from lack of modularity, imprecise performance, and relatively small potential for educational use.

Methodology

Overview

The main purpose of the work reported in this paper was to evaluate the suitability of PS methodology for representing knowledge about neurological localization. The emphasis was on how well the PS approach to automated localization would express the necessary problem-solving knowledge involved. As such, this project forms part of a more comprehensive study* into diagnosis in neurology and into how techniques in AI can be used to capture this diagnostic process in a computer program.

A MYCIN-like rule-based program that localizes damage to the central nervous system in an unconscious patient was designed and implemented. The program interprets a neurological examination in the sense that it classifies comatose patients into the four categories described in the preceding section. The localization problem was limited to unconscious patients in order to keep the project to a reasonable size: I was more interested in the principles involved and in the representational adequacy of PS's than in producing a large, comprehensive system.

The program's performance during an interactive session is illustrated below, followed by a description of how it produces its results. The results, and what was learned from them, will then be discussed.

Example of an Interactive Session

An actual interactive session with the program is shown below. The program's output is in lower case letters and the user's responses are in upper case.

Ready:
NEWCASE.

Describe patient's posturing:
ABSENT BILATERALLY.

Describe patient's pupils:
THE RIGHT IS DILATED-FIXED AND THE LEFT IS NORMAL.

Describe patient's respirations.
NORMAL.

Describe patient's eye-position:
STRAIGHT BILATERALLY.

Describe patient's left-calorics:
THE LEFT EYE DEVIATES-OUT AND THE RIGHT EYE DEVIATES-IN.

*The NEUREX Project (Reggia, 1978).

Describe patient's right-calorics:
THE RIGHT EYE DEVIATES-OUT AND LEFT DEVIATES-IN.

Describe patient's corneals:
THIS IS GARBAGE.

The word garbage is unrecognized in the current context. Description ignored.
(Options are: present, absent, untested)

Describe patient's corneals:
PRESENT BILATERALLY.

Describe patient's grimace:
LEFT ABSENT BUT RIGHT IS PRESENT.

Describe patient's tone:
NORMAL.

Describe patient's DTR's:
LEFT HYPERREFLEXIA BUT NORMAL ON THE RIGHT.

Describe patient's plantar-responses:
EXTENSOR ON THE LEFT AND FLEXOR ON THE RIGHT.

Assessment:
    Right supratentorial lesion with certainty of 0.90
    Patient is herniating with certainty of 0.54.

The interactive session thus consists of a series of program-generated questions to which the user responds by supplying the relevant information about the patient's examination. While the vocabulary available to answer questions is limited, it is basically familiar to the clinician.

If the program ever encounters a word that it does not recognize, it ignores the response, offers options from which the user may choose, and then repeats the question. This is illustrated above.

Two other features of the program not shown here for the sake of brevity are a trace that prints out the result as each rule is evaluated, and a display of the database that describes the patient and can be printed at the user's request.

Database

"Dynamic knowledge" about the patient is kept in the database and consists of a set of attributes possessed by the patient. These are of two types: *examination attributes* (e.g., REFLEXES, TONE, PUPILS, RESPIRA-TIONS) that are supplied by the user, and *inferred attributes* (e.g., FOCAL-LESIONS). Each attribute can take on one or perhaps several *values* on one or both *sides*. The patient is thus represented as a collection of *attribute-value-side triplets* that have been chosen to correspond to the way that clinical neurologists characterize unconscious patients. Associated with each triplet is information indicating why the system believes that fact to be true (a list of the rules it used to infer an attribute's value) and estimates of its degree of belief that the information in the triplet is correct. Some examples of inferences that the program can make about a patient are: whether various cranial nerve lesions are present or absent, whether the corticospinal tract or medial longitudinal fasciculus is damaged, and whether the pons and/or the midbrain are intact.

Rule-Based Knowledge

During the development of the rules representing neurological localization knowledge, it was convenient to organize them using a "natural inference hierarchy" (see Figure 3). Successively higher levels in this hierarchy represent greater and greater levels of abstraction of the information about the patient. The rules themselves are syntactically similar to those of MYCIN and their general encoded form is:

> (⟨rule number⟩   (antecedent clause list)
>   (consequent clause list))

Consider an example rule (RULE 48) in its LISP-encoded form:

(R48 (AND (UNILATERAL FOCAL-FINDINGS CN7 SIDE)
          (UNILATERAL FOCAL-FINDINGS CST
          CONTRALATERAL))
     ((CONCLUDE FOCAL-LESIONS PONS SIDE TALLY 0.9)))

This rule expresses the belief that:

> *If* the patient has a seventh nerve palsy on one side
>   *and* corticospinal tract findings on the contralateral side
> *then* there is strong evidence (0.9) that the patient has a focal lesion in the
>   pons on the side of the seventh nerve palsy.

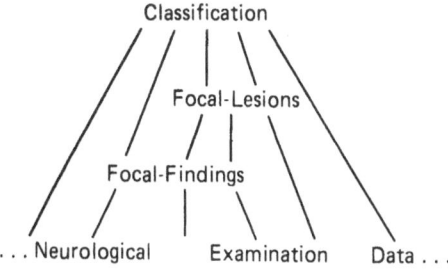

Figure 3. Natural inference hierarchy.

In this simple rule, AND and UNILATERAL are predicates,* FOCAL-FINDINGS and FOCAL-LESIONS are attributes, and CST, CN7, and PONS are the attribute values. TALLY is a variable whose value will reflect the "certainty" with which the conclusions in the rule can be inferred, given that the antecedents are true, while SIDE and CONTRA-LATERAL are dependent variables (if SIDE = LEFT, then CONTRA-LATERAL = RIGHT, and vice versa). The predicate UNILATERAL is an example of a domain-specific "macropredicate," a concept which will be discussed below.

All rules were hand-coded. As they are translated into an internal, program manipulatable format, they are screened both for syntactic and semantic errors. For example, the value of the attribute of each clause is checked to make sure it *is* a possible value for that attribute. Such simple screening significantly decreased development time for the program.

The Rule Interpreter

The control structure used is consequent-driven and produces an implicit depth-first exhaustive search of an AND/OR goal tree (Nilsson, 1971). The recursive nature of the control flow involves:

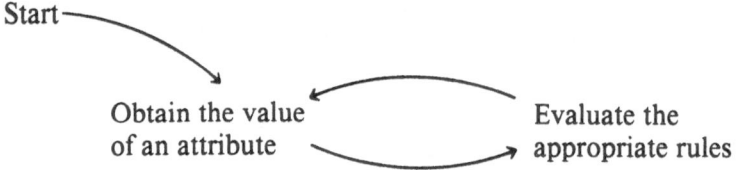

Informally, the program begins with the start goal, "In which class of coma does the patient belong?" and then sets up subgoals such as "What focal lesions does the patient have?" These, in turn, may set up more subgoals or result in the program's requesting information about the patient from the physician. The probabilities or "certainty factors" associated with each inference are propagated up the goal tree as in MYCIN (Shortliffe and Buchanan, 1975) and appear in the assessment printed at the end of an interactive session.

User Interface

When the program attempts to obtain the value of an examination attribute, it discovers that there are no rules which it can apply. It therefore calls a simple IO routine that asks the user for the information and then transforms the response into a standard form by rearranging words,

*The word "predicate" is used loosely here in that the functions involved do not take on the value of T or F but rather assume a numerical value reflecting their probability of being T or F. The word is used in a similar fashion throughout this paper.

replacing synonyms, etc. A primitive keyword-based approach is used. The resultant attribute-value-side triplets are then stored in the database.

Implementation

The program was implemented in Wisconsin LISP and run on the Univac 1108 at the University of Maryland.

Performance and Testing

The program was tested on several simulated patients in each of the four categories and in combinations of the categories until it performed at a level of competence similar to that demonstrated in the interactive session illustrated earlier.* A significant amount of development time was spent correcting errors or omissions in the rules that were revealed during test sessions. Of course, this in no way formally verified or validated the program or its knowledge base, but it did produce a program that worked reasonably well in many cases.

Implementation of the control program (rule interpreter) and database was straightforward. The program was easy to use and essentially no waiting was required for computer responses.

Representation of Neurological Localization Knowledge

The main purpose of this study was to evaluate the generality of MYCIN-like PS's for medical problem solving by applying them to a new medical domain. The experience obtained during the development of the program described above has led to the following observations and conclusions, some of which are in disagreement with opinions that have previously been expressed in the literature.

1. Expressing neurological localization knowledge as a collection of rules is very difficult.

For the limited domain chosen, 97 rules are currently used to represent neurological localization knowledge. Organizing and writing even this limited set of productions was very difficult. (The reasons for this are given in the following two sections.) On two occasions, entire collections of rules were discarded before the current set was adopted. This difficulty in expressing knowledge as a set of rules is not a problem isolated to the current study. It has been encountered fairly consistently by those com-

---

*The assessment given there is reasonable.

posing both medical and nonmedical rule-based programs (e.g., Shortliffe, 1976, p. 85 and 157; Buchanan, et al., 1970).

2. A collection of rules is not a good model of the organization of neuro-logical localization knowledge as used by the physician; that is one reason that their creation is so difficult.

It has often been felt that the difficulty in acquiring rules reflects a prob-lem of information transfer between a domain-knowledgeable person and a computer-scientist. While such communication is indeed a formidable task, this problem alone cannot explain the difficulties encountered in creating rules, because the same difficulties occur even when both the rule-provider and the computer-scientist are the same individual (as occurred here and with PECOS; see Barstow, 1977a).

In the neurological domain, much of the problem arose from the fact that a collection of rules is simply not a good model of how a physician organizes the knowledge that he uses to localize nervous system damage. Much of this information is arranged in a framelike fashion (Minsky, 1975) with a great deal of interdependence between individual facts. Thus, although logically possible to a certain extent, reformulating it into discrete rules is very difficult. In addition, both introspection and discus-sion with other clinicians have convinced me that a conceptual, visually-oriented representation of the nervous system is used by physicians in localizing damage to the nervous system.* Capturing knowledge that is organized in such a fashion as a set of rules would appear to be very difficult.

While the assertion that neurological knowledge is not really organized as a set of rules may seem intuitively obvious to many clinicians, the subject is more complex than it first appears and is related to important theoretical questions in artificial intelligence. Some cognitive psycholo-gists exploring the use of "pure" PS's to model human thinking feel that the success of PS's at this task arises out of their similarity to actual human cognitive processes (e.g., Newell and Simon, 1972, pp. 803–806). However, the difficulty in acquiring the necessary rules for neurological localization and for other domains would appear to run counter to this belief. While it could be suggested that such a rule-based arrangement of knowledge in the human mind might be inaccessible to introspection, such an assertion remains to be proven.

It has also been suggested that the *good performance* of PS's supports their role as models of fundamental underlying cognitive mechanisms.

---

*This "internal image" is acquired by studying neuroanatomy and neurophysiology during one's medical training. Kleinmuntz, a psychologist who has used simulated patient encoun-ters to investigate the cognitive processes used by neurologists in reaching a diagnosis, has arrived at a similar conclusion (Kleinmuntz, 1968).

This argument must be interpreted with caution. Correspondence between the *external* behavior of a model and that of the actual system being simulated (here a computer program and human cognitive mechanisms respectively) is a necessary but not sufficient condition for correspondence between their *internal* mechanisms. If good performance alone provided strong support for the role of a programming methodology as a model of human cognition, then even purely statistical diagnostic programs, having been shown in some cases to perform as accurately as senior physicians (e.g., Dombal, 1975; Williams and Stern, 1975) and programs like ELIZA (Weizenbaum, 1965) might be offered as simulations of human thought processes.

3. The interpretation of neurological examination abnormalities is highly context-dependent. This leads to combinatorial problems.

Because individual rules lack a global perspective of the problems being solved, one must include all the necessary context for their application in the antecedent clauses (Davis, et al., 1977). In the domain of neurological localization, this leads to some significant problems. For one thing, all of the relevant factors for applying a rule are not always obvious and this results in errors in the rules. But even more important, the interpretation of examination abnormalities can be very context-dependent at times. For example, the "meaning" of a unilateral dilated and fixed pupil depends on whether long-tract findings are present, whether a patient is awake or unconscious, whether the neck is supple or rigid, etc. If one attempts to incorporate all relevant context into a rule's antecedent clauses there is a rapid growth of the number of rules present. The problem of "combinatorial explosion" was avoided in the present study by limiting the task to evaluation of unconscious patients.

4. One way to improve the compactness and understandability of rules is to create and use domain-specific "macropredicates."

In one of the earlier sets of neurological localization rules that was tried and abandoned, an attempt was made to use the same predicates as were used in MYCIN (Shortliffe, 1976, pp. 101–108). These predicates, such as KNOWN, MIGHTBE, SAME, NOTSAME, etc. led to large and awkwardly expressed rules.

This difficulty was overcome by rewriting the rules using *"macropredicates"* that are each a combination of more primitive predicates similar to those used in MYCIN. For example, the predicate UNILATERAL which was illustrated above has special significance in neurological localization and forms a natural concept for a clinician. The predicate UNILATERAL is true if an abnormality is present on one side *and* if that same abnormality is *not* present on the opposite side. It thus expands (at run times) into more primitive predicates similar to those of MYCIN:

(UNILATERAL ⟨attribute⟩ ⟨value⟩ ⟨side⟩)
⇓
(AND (HAS ⟨attribute⟩ ⟨value⟩ ⟨side⟩)
        (NOT (HAS ⟨attribute⟩ ⟨value⟩ ⟨opposite side⟩))))

Expressing concepts in a form that is familiar to physicians is important not only for the person providing rules but also for their use in an educational setting where a program "explains" its decisions by demonstrating the rules it used in making them. Those interested in constructing general-purpose rule interpreters that can be applied to main domains by simply changing the set of rules used by the program would probably find it helpful to include provisions for the user to define such domain-specific macropredicates at will.

Conclusion

When compared to more conventional programming methods (such as branching-logic, numerical formulas, and purely statistical calculations), PS methodology provides some significant advantages in terms of modularity and educational potential. As illustrated in Figure 1, this has led to many ongoing attempts to implement PS's that solve real-world problems.

The study reported here has provided a critical evaluation of the generality of MYCIN-like PS's in medicine by implementing one for a relatively small neurological localization problem. Attempts to reformulate the relevant knowledge into rules revealed several problems that should serve as a guide for further research. One fruitful approach might be the combination of PS methodology with other ways of representing knowledge, something which is being tried in PROSPECTOR (Duda, et al., 1977). Alternatively, one might adopt a more frame-like organization of information. This latter approach will guide the author's future search for improved representation of neurological knowledge.

Acknowledgment. This research was funded by an NIH Teacher-Investigator Development Award (KO7 NS 1044-78) from the NINCDS. Their support and encouragement are gratefully acknowledged. This report benefited in large measure from the constructive criticism offered by Chuck Rieger, Erland Nelson, Ramesh Khurana, Phil London, Steve Small, and Milt Grinberg of the University of Maryland.

Appendix
Additional Example Rules

```
(R11 (AND (UNILATERAL POSTURING DECEREBRATE SIDE)
          (UNILATERAL POSTURING DECORTICATE CONTRALATERAL))
     ((CONCLUDE FOCAL-LESIONS CEREBRAL-HEMISPHERE CONTRALATERAL TALLY
       0.8)
      (CONCLUDE CLASSIFICATION HERNIATING BILATERAL TALLY 0.8)))

(R22 (AND (BILATERAL GRIMACE PRESENT)
          (UNILATERAL CORNEALS ABSENT SIDE))
     ((CONCLUDE FOCAL-FINDINGS CN5 SIDE TALLY 0.7)))

(R58 (AND (BILATERAL FOCAL-FINDINGS MLF)
          (BILATERAL FOCAL-FINDINGS CST))
     ((CONCLUDE FOCAL-LESIONS PONS BILATERAL TALLY 0.8)))

(R69 (AND (UNILATERAL FOCAL-FINDINGS CN3 SIDE)
          (ANY FOCAL-LESIONS CEREBRAL-HEMISPHERE))
     ((CONCLUDE CLASSIFICATION HERNIATING BILATERAL TALLY 0.6)))
```

References

Barstow, D.: Automatic Construction of Algorithms and Data Structures Using a Knowledge Base of Programming Rules, *Memo AIM-308,* Stanford Artificial Intelligence Laboratory, 1977a.

Barstow, D.: A Knowledge-Based System for Automatic Program Construction, *Proc. 5th IJCAI,* 1977b, 382–388.

Buchanan, B., Sutherland, G., and Feigenbaum, E.: Rediscovering Some Problems of Artificial Intelligence in the Context of Organic Chemistry, Machine Intelligence, 5, Meltzer and Michie (eds.), Edinburgh University Press, 1970.

Davis, R. and King, J.: An Overview of Production Systems, Machine Intelligence, 8, Elcock and Michie (eds.), 1976, 300–332.

Davis, R., Buchanan, B., and Shortliffe, E.: Production Rules as a Representation for a Knowledge-Based Consultation Program, *Artificial Intelligence,* 8, 1977, 15–45.

Dombal, F.: Computer Assisted Diagnosis of Abdominal Pain, in Advances in Medical Computing, Rose and Mitchell (eds.), Churchill-Livingston, N.Y., 1975, 10–19.

Duda, R., Hart, P., Nilsson, N., et al: *Development of a Computer-Based Consultant for Mineral Exploration,* October 1977 (Annual Report), Stanford Research Institute.

Hayes-Roth, F., Waterman, D., and Lenat, D.: Principles of Pattern-Directed Inference Systems, in Pattern-Directed Inference Systems, Waterman and Hayes-Roth (eds.), Academic Press, 1978, 577–601.

Haymaker, W.: Bing's Local Diagnosis in Neurological Diseases, C.V. Mosby Co., 1969.

Kleinmuntz, B.: The Processing of Clinical Information by Man and Machine, in

Formal Representations of Human Judgment, B. Kleinmuntz (ed.), John Wiley and Sons, 1968, 149–186.

Martin, Friedland, King, and Stefik: Knowledge Base Management for Experimental Planning in Molecular Genetics, *Proc. 5th IJCAI,* 1977, 882–887.

Meyer, A. and Weissman, W.: Localization of Lesions in the Human Nervous System by Computer Analysis, *IEEE Trans. Biomed. Eng.,* EME-20, 1973a, 194–200.

Meyer, A. and Weissman, W.: Computer Analysis of the Clinical Neurological Examination, *Comp. Biol. Med.,* 3, 1973b, 111–117.

Minsky, M. A Framework for Representing Knowledge in The Psychology of Computer Vision, P. Winston (ed.), McGraw-Hill, 1975.

Newell, A. and Simon, H.: Human Problem Solving, Prentice-Hall, 1972.

Nii, H. and Feigenbaum, E.: Rule-Based Understanding of Signals, in Pattern-Directed Inference Systems, op. cit.

Nilsson, N.: Problem-Solving Methods in Artificial Intelligence, McGraw-Hill, 1971.

Plum, F. and Posner, J.: Diagnosis of Stupor and Coma, F.A. Davis Co., 1972.

Reggia, J.: Technical Report, Department of Computer Science, University of Maryland, 1978.

Shortliffe, E.: *Computer-Based Medical Consultations:* MYCIN, Elsevier, 1976.

Shortliffe and Buchanan: A Model of Inexact Reasoning in Medicine, *Math. Biosciences,* 23, 1975, 351.

Stewart, A. and Cala, L.: Mathematical Method to Utilize a Computer for Diagnosis of Site and Type of Intracerebral Mass Lesions, *Brit. J. Radiol.,* 48, 1975, 97–100.

Vinken, P. and Bruyn, G., editors: Localization in Clinical Neurology, Vol. 2 of Handbook of Clinical Neurology, North-Holland Publishing Co., 1967.

Virnich, H. et al.: CEDI—ein Computerprogramm zur praeoperatien Art-diagnostik hirnorganischer Prozesse, *Meth. Inform. Med.,* 14, 1975, 19–25.

Walser, R. and McCormick, B.: A System for Priming a Clinical Knowledge Base, *Proc. Nat. Comput. Conf.,* 1977, 301–307.

Waterman, D. and Hayes-Roth, F.: An Overview of Pattern-Directed Inference Systems, in Pattern-Directed Inference Systems, op. cit.

Weiss, S., Kulikowski, C., and Safir, A.: Glaucoma Consultation by Computer, *Computers in Biol. Med.,* 8, 1978, 25–40.

Weizenbaum, J.: ELIZA—A Computer Program for the Study of National Language Communication Between Man and Machine, *CACM,* 9, 1965.

Williams, R. and Stern, R.: Computer Diagnosis of Jaundice, in *Advances in Medical Computing,* J. Rose and J. Mitchell (eds.), Churchill-Livingston, 1975, 20–27.

18

Rule-Based Drug Prescribing Review: An Operational System

Stuart M. Speedie, Francis B. Palumbo,
David A. Knapp, and Robert Beardsley

This paper describes an operational computer-based system for conducting Drug Prescribing Review (DPR). The system was designed to: arrive at specific judgements about the potential problems of drug orders taking into account the characteristics of the patient and their medical conditions; be relatively independent of the structure of the data base; and be capable of expressing and evaluating any DPR criteria. This was accomplished by developing a hierarchical, rule-based system for expressing and evaluating DPR criteria. This system was then linked with a specific patient data base for implementation. Drug orders for 65 patients have been evaluated and the results agree with expert judgement. The rule-based DPR system appears to be a feasible method of evaluation of drug orders.

Introduction

The drug use process encompasses prescribing, dispensing and administration of drugs. The importance of the drug use process in health care is evidenced by the fact that 2 of every 3 outpatient visits result in at least one prescription order and that the average American obtains over 5 prescription orders a year [1]. Thus drug utilization review, the methodology which evaluates the quality of the drug use process, is an important component of any health program's quality assurance effort. Drug Prescribing Review (DPR) is currently a major area of activity within the larger area of drug utilization review. DPR focuses on the quality of physician decision-making in prescribing drug therapy. Ideally, DPR evaluates the appropriateness of the drug, the dose, the regimen and other prescription parameters in terms of the individual patient and their total drug therapy.

© 1981 by the Institute of Electrical and Electronics Engineers, Inc. Reprinted, with permission, from Proceedings of the Fifth Annual Symposium on Computer Applications in Medical Care, Washington, D.C., November 1–4, pp. 598–602.

The goal of DPR is to improve the quality of the physician's decision-making through examination of prescribing patterns and the development and implementation of suitable optimizing actions.

DPR is not a single process, but rather a continuum of processes which vary in complexity, exhaustiveness and finality of judgement. At one end of the continuum is in-depth review which takes into account the characteristics of the drug therapy and the patient in order to arrive at a clinical judgement as to the appropriateness or inappropriateness of the specific drug order. In-depth review in DPR is a comprehensive, time consuming process which results in a definitive decision for a given drug order for a given patient.

Toward the other end of the continuum is screening review. In contrast to in-depth review, a screening review is not intended to arrive at a definitive judgement about a particular drug order. Rather, it attempts to identify drug orders which are *potentially* inappropriate. Screening review is a "gross" measure of prescribing quality which relies on general guidelines for appropriate prescribing rather than on a final decision using professional clinical judgement. Obviously, screening is a simpler and less exhaustive process than in-depth review. However, screening review has demonstrated its usefulness for evaluating programs which focus on prescribing patterns of physicians [2,3].

The development of "optimal" drug prescribing guidelines or criteria is a critical activity in screening type DPR. Criteria development is a multi-stage process involving a thorough literature search and multiple draft-review-revision cycles. Ideally, these criteria should specify: (a) appropriate indications, (b) exceptional conditions, (c) acceptable dosage forms and ranges, (d) appropriate regimens, (e) required contraindicated and concomitant therapy, (f) required monitoring, and (g) the impact of relevant patient variables [4].

Unfortunately, the automated DPR systems such as that described by Helling [5] most often have *not* incorporated *all* the desirable components of criteria. Judgement concerning patient characteristics as it relates to appropriate indications and other relevant patient variables is either left to "professional judgement" or not included at all. The exception to this statement is the work reported by Maronde [6] in which some diagnostic information was included in their automated DPR system. Even that system, however, did not take advantage of the full range of information available in the patient record. The challenge, then, is to develop a computer-based DPR system which incorporates these patient factors into the criteria, in order to more closely approach the ideal DPR system.

It should be evident from this discussion that the criteria are the core of any DPR system. Their formulation and implementation are the key to system quality. Thus any computer-based system must develop procedures which are capable of evaluating all the components of ideal criteria.

The Rule-Based DPR System

The design goals for the computer-based DPR system reported in this paper focused on specificity, independence and generalizability. The system should be:

1. Able to arrive at a specific judgement about the potential problems in a drug order by applying DPR criteria.
2. Independent of the structure of the patient data base.
3. Independent of the specific considerations for any particular drug.
4. Capable of expressing any criteria which may be used in DPR.

The method chosen to achieve these goals was the creation of a pattern directed inference system as described by Waterman & Hayes-Roth [7]. Furthermore, the criteria were defined as components of a rule-based system; a pattern directed inference system where the system consists of a set of logical expressions called "rules." Due to the relatively complex nature of DPR criteria, it was necessary to use a hierarchy of "rules" to simplify the process of criteria building. The outcome of this process has been a hierarchical rule-based system for evaluation of drug orders with respect to DPR criteria. In a sense, the resulting system is a meta-language in that there is great flexibility in expressing criteria, as long as acceptable elements are combined according to the syntactical rules of symbolic logic and set logic.

The hierarchical nature of the rule-based system derives from the nature of DPR criteria. In the purest sense, each level of the hierarchy is a set of "rules." The lowest level rules are "assertions" or simple testable factual statements about information in the patient data base. The next level consists of rules in the more conventional sense in that they combine assertions using the operators of symbolic logic and set theory. The third level consists of "screens" which logically combine rules in order to evaluate the appropriateness of specific drug entities. These screens are triggered by the appearance of the corresponding drug entity in a physician's order.

The project with which this system is associated has developed a set of conventional screening criteria for digoxin, analgesics, sedative/hypnotics, antianxiety agents, diuretics and laxatives covering approximately 300 different products. These criteria establish the base from which the computer-based DPR system was developed. In the following sections, the development and derivation of assertions, rules and screens will be described.

Criteria Implementation

The first step in implementing these DPR criteria was to reorganize them into a set of "screens" for each generic drug entity or combination product. These screens correspond to the sections of the criteria, and have the following labels:

1. *Unacceptable per se*—indicates that a particular drug is not acceptable for a particular indication.
2. *Do not evaluate*—identifies exceptional conditions under which an order is not evaluated. (e.g. the use of aspirin is not evaluated when the patient has rheumatoid arthritis).
3. *Indications*—determines whether or not the appropriate indications are present for the use of a particular drug.
4. *Dosage*—determines whether or not the dosage is within the recommended range.
5. *Regimen*—determines if the regimen used in a drug order is one of those which is acceptable for this drug.
6. *Contraindicated Conditions*—checks for the existence of contraindicated conditions for a particular drug.
7. *Contraindicated Drugs*—determines if the patient is taking any drugs which are contraindicated with the drug being evaluated.
8. *Required Concomitant Drugs*—checks the active drug orders to determine whether or not any required additional drugs have been ordered.
9. *Required Monitoring*—checks for the existence of the various types of monitoring which may be required for a particular drug.

The organization of the criteria into this format allows the construction of a computer-based DPR system which uses the prescribed drug as its starting point. It associates each drug of interest with a set of screens and permits evaluation of drugs with respect to these screens using information from the patient data base. The following example illustrates a typical screen.

> *Example:* Digoxin dosage screen: The order passes the screen if:
> *the indication is atrial fibrillation AND the dose is between .125 mg and .75 mg OR the indication is other than atrial fibrillation AND the dose is less than or equal to .5 mg.*

A process of logical decomposition was employed to derive rules and assertions from the screens. Though the original analysis worked from screens to assertions, they are best described in the reverse order.

Assertions are the first and lowest level in the hierarchy. They are simple statements of fact which may be directly evaluated. They are constructed as follows:

SUBJECT RELATION PREDICATE PREDICATE-MODIFIER

> SUBJECT is defined as one of:
> A. Any data item from any entry in the patient data base for a particular patient.

B. A data item from a drug order being reviewed.

RELATION is defined as one of:

A. =, or "same as"

B. less than,

C. less than or equal to,

D. greater than,

E. greater than or equal to.

PREDICATE is defined as one of:

A. A data item from the drug order being reviewed

B. A "constant" (i.e., a specific value such as CHF).

C. A "constant" list,

D. Any data item from any entry in the patient data base.

PREDICATE-MODIFIER is defined as:

A. a value which is added to the predicate when the predicate refers to a date.

Assertions operate on individual data entries in the patient data base. The evaluation of these assertions employs set logic. That is, an assertion such as "Problem = Hypertension" finds the set of all entries for which the problem is "Hypertension." Thus the outcome for an evaluation of an assertion is a list or set of entries for which a particular condition defined by the assertion is true. Associated with this list is a truth value. It is "True" if there is at least one entry on the list. Otherwise, the truth value is "False."

Subrules are rule components which combine or operate on assertions. They use the set operations of Intersection, Union, and Complement. These operations are needed to identify entries which satisfy multiple conditions or assertions. For example, a subrule might be "Problem = Renal Failure Intersect Problem = CHF." The outcome of the evaluation of this subrule is a list of data entries which satisfy both assertions. That is, it will select a set of entries which indicate the patient has both renal failure and CHF. Subrules are also used to combine the truth value of assertions and/or subrules using the logical operators AND, OR and NOT.

Similarly, rules are logical combinations of subrules and assertions which encompass a specific topic such as dosage or regimen. Rules may have a range of complexity from a single assertion to combinations involving long lists of subrules and assertions. They also take on truth values of True or False.

Screens are logical combinations of rules using the logical operators AND, OR and NOT. The screens are evaluated based on the results of the rules. The truth value of the rules are used and combined in several ways in order to determine the truth value of the screen. If all rules are satisfied, then the drug order passes the screen, and the value is True; otherwise, it does not and the value is False. Again, screens can range in complexity as do rules. They may range from a long list of rules which are combined using logical operators to a simple restatement of a particular rule.

As can be seen from the discussion of this hierarchical rule based system, the process of evaluation must begin at the lowest level and work up. That is, when a particular drug order is encountered, the assertions related to that generic drug entity are evaluated first. The assertions are then combined using the subrules to yield a True or False value for a particular rule. After all rules are evaluated, the screens are evaluated to determine whether or not the particular drug order passes or fails the screens.

This approach to implementing the DPR criteria was successful for almost all of the criteria developed by the project. However, there were two types of rules which were not amenable to this type of analysis. As a result, two special rules and means for evaluating them were developed.

The first of these rules concerned related monitoring which must take place within given time intervals. For example, when digoxin is being used, the BUN should be monitored every six months. Furthermore, if this test results in an abnormal value, the appropriate action should be taken and the test should be repeated within thirty days. Logical expressions cannot accurately represent this situation since it is a time dependent, outcome dependent judgment. Therefore, a special routine was developed. This special rule is triggered when a drug is encountered that requires periodic monitoring—e.g., a diuretic. The rule checks the dates of the appropriate monitoring test to see if they fall within the acceptable limits. Furthermore, it checks the value of the previous test and evaluates the time lapse based on that. The rule identifies all those tests which pass or satisfy this special rule and those which do not and, furthermore, returns the value for the rule which indicates whether or not the monitoring was performed satisfactorily.

The second rule involves the case where a specific action is required, based on the results of monitoring. For example, when a patient is on a diuretic, and the potassium level is abnormally low, a potassium supplement should be ordered within 48 hours or the existing supplement should be increased. This rule is designed to check any of the following when an abnormal test is encountered:

1. a new drug was ordered.
2. the dose of a current drug was increased.
3. the dose of a current drug was decreased.
4. a current drug was discontinued.
5. a current drug was changed to a different drug in the same class.
6. another drug was added from the same class.

This rule also checks to see if the appropriate action was taken within the specified period of time. Again, a result of True or False is returned for this rule.

DPR Implementation

The principal components of the software system which implements this DPR system are a patient data base, a DPR criteria data base, and a DPR program. The following sections will describe each component.

Patient Data Base

The patient data base is a hierarchically structured set of files that are currently implemented using the Hewlett Packard IMAGE/3000 data base language. The data base is organized into 2 master files (PATIENT and DOCTOR) and 3 detail files (PT-VAR-DEMO, PT-COND and DRUG-ORDERS). The PATIENT file contains identifying demographic for each patient including project identification number, age, sex. The PT-VAR-DEMO file contains information on the patient's marital status and physicians. The PT-COND file contains information on the patients' conditions including patient status, dietary restrictions, drug sensitivities, medical problems (chronic & acute), laboratory tests, weights and blood pressures. DRUG-ORDERS contain all drug orders for the patients. The PATIENT file is linked to all the others via the project identification number. The DOCTOR file of physician demographic information is linked to the DRUG-ORDERS file by the physician identification number. Thus the DRUG-ORDERS file may be accessed either by patient identification number or physician identification number.

Criteria Data Base and DPR Program

The purpose of these components is to carry out the evaluation of selected drug orders using the criteria developed for the prescribed drug entities. As the system is designed, it is a general rule evaluation system which can be used to evaluate any drug entity against any set of criteria. This system makes use of a DPR criteria data base which contains all the necessary information to evaluate selected drug orders. It is organized into five independent files of information. As described previously all drug orders are evaluated with respect to several screens. In order to accommodate this, the data base is initially organized around classes of drugs. For each class of drugs there is a Dictionary File which relates information in the criteria to information contained in the various files of the patient data base. The purpose of the dictionary is to increase the independence of the DPR system from the patient data base. All references to items from the patient data base by criteria are essentially references to the dictionary which in turn points to the appropriate item in the data base. As a consequence, the data base can be changed or modified or an entirely new data base created and only the dictionary need be changed rather than the whole criteria system. Within each drug class, there are a variety of screens which correspond to the major criteria for indications, dosage, regimen, etc. The information describing these screens is contained in the Screens File. The rules which the screens reference and combine are contained in the Rules File. The entries in the Rules File are sub-rules or assertions.

Since each prescription is written for a particular drug entity rather than a generic class of drugs, it is necessary to have some mechanism for selecting the appropriate dictionary and set of screens and rules. This is accomplished using the Generic Drug File. All generic drug entities, as well as combination products are coded as entries in the Generic Drug File. This file contains the identification number of the generic drug entity, a brief description of the drug, the number of the dictionary for this drug, and a list of up to 25 applicable screens for the selected study drugs. If a drug is not one of the study drugs, it is so identified.

The DPR program which utilizes this criteria data operates by selecting a patient from the Patient File. The system retrieves all the patient conditions for that patient as well as all drug orders. The system then indexes through each drug order in the following manner.

The generic drug identification number is retrieved from the drug order, and that is used to retrieve the appropriate record from the Generic Drug File. This information is used to retrieve the appropriate dictionary from the Dictionary File and the relevant screens from the Screens File. The information in the screens is used to retrieve rules from the Rules File. The system then evaluates each of the assertions or sub-rules, and then each of the rules. This information is used to evaluate each of the screens. The information which results from each screen evaluation is then placed into a data entry item for that drug order. The result is a series of fields in the drug order which indicate, among other things, whether or not an acceptable indication was present, whether or not the drug dosage fell within the acceptable range as well as all information about all the other aspects of the criteria. At this point the system goes on to the next drug order and repeats the process. At the completion of all drug orders, the system returns the evaluated drug orders to the patient data base and indexes on to the next patient. It proceeds in this manner until all orders for all patients are evaluated, at which time a review cycle is completed.

Results of Implementation

The system has been applied to 65 patients from five nursing homes. Medical records information and all drug orders were collected for a three month period prior to March 18, 1981. This resulted in a total of 657 drug orders. Applicable criteria had been developed for 312 of these orders and they were evaluated for potential problems using DPR screens. The

Table 1. Summary of Potential Problems Reported for a Sample of 65 Patients

Class Group	N	NA	Ind.	Dose	Reg.	CC	CD	RM	RD	RA	NE
Digoxin	28	0	8	0	0	0	0	56	0	2	0
Sedative/Hyp.	28	0	14	7	8	0	0	0	0	0	0
Analgesics	78	7	0	3	14	0	1	0	0	0	21
Diuretics	41	0	18	1	0	0	0	60	0	2	0
Laxatives	137	34	79	6	26	8	7	0	0	0	0
Total	312	41	119	17	48	8	8	116	0	4	21
Total Possible		312	312	312	312	-	-	345	-	-	312
Percentage		13%	38%	5%	15%	-	-	37%	-	-	7%

KEY: NA–Not Acceptable for use in the elderly; Ind–Indications not found in the patient's record; Dose–Dosage outside recommended range; Reg.–Regimen not one of those recommended; CC–patient has a Contraindicated Condition; CD–patient was also prescribed a Contraindicated Drug; RM–required monitoring not found in the patient's record; RD–required concomitant drug not found; NE–Not Evaluated due to exceptional conditions; "-"–data not available at this time.

results of this evaluation are detailed in Table 1. In order to check the accuracy of these results, all patient drug orders were manually evaluated using the DPR criteria and the results were confirmed. Thus the system appears to be operating properly.

These early results reveal that the majority of potential problems are due to missing indications or lack of appropriate monitoring procedures. Since a majority of problems with indications occur in the prescribing of laxatives, a question may be raised about the appropriateness of the indication criterion for laxatives. The flexibility of the system is demonstrated by the fact that if it was decided that no indication should be required for laxatives, only a simple change in the criteria base is needed. The indication screen would be deleted from each of the laxatives in the Generic Drug File.

The preliminary results indicate that system has met the major design goals. It is capable of expressing and evaluating all criteria developed so far. The elements of the patient data base and the criteria data base may be modified easily. The Dictionary File allows independence of the criteria and the patient data base.

Summary

The rule-based DPR system appears to be a feasible method of automated evaluation of drug orders. It does detect potential problems by applying screening criteria. The language-like nature of the criteria makes the system both flexible and independent of the specific structure of the patient data base. Such a system should be capable of evaluating screening criteria for most drug entities.

Acknowledgment. This research was supported by grant no. 5 RO1 HS 03305-03 from the National Center for Health Services Research.

References

[1] Stolley, P.D. and Lasagna, L., "Prescribing patterns of physicians," *J. Chronic Dis., 22,* 395(1969).

[2] Knapp, D.A., Brandon, B.M., Knapp, D.E., Klein, L.S., Palumbo, F.B. and Rohit Shah, "Incorporating diagnosis information into a manual drug use review system," *Am. J. Pharm. Assoc.,* NS17, 103(February 1977).

[3] Brandon, B.M., Knapp, D.A., Klein, L.S. and John Gregory, "Drug usage screening criteria," *Am. J. Hosp. Pharm., 34*(2), (February 1977).

[4] Knapp, D.A., Brandon, B.M., and West, S., "Development and application of criteria in drug use review programs," *Am. J. Hosp. Pharm., 31*(7), 648(July 1974).

[5] Helling, D.K. Hepler, C.D. and Herman, R.A., "Comparison of computer-assisted medical record audit with other drug use review methods," *Am. J. Hosp. Pharm., 36,* 1665(December 1979).

[6] Maronde, R. Drug Utilization Review with On-Line Computer Capability, Social Security Administration, Washington, D.C., 1973.

[7] Waterman, D.A. and Hayes-Roth, F. Pattern Directed Inference Systems, New York: Academic Press, 1978.

A Heuristic Approach to Risk Analysis in Computer-Assisted Medical Management

Perry L. Miller

The paper describes the heuristic approach to risk analysis taken by ATTENDING, an Artificial Intelligence system designed to critique a physician's plan of anesthetic management. To critique a physician's plan, the system must know about the different risks and risk tradeoffs involved in a patient's management. It must also be able to manipulate this knowledge in a flexible way. The paper discusses the three principles on which ATTENDING's heuristic approach to risk is based, and also describes how each principle is currently implemented.

1. Introduction

The ATTENDING system [1] is being developed to critique a physician's plan for anesthetic management. To use the system, an anesthetist describes a patient's medical problems together with a tentative anesthetic plan. ATTENDING then critiques the plan from the standpoint of the underlying risks involved in that patient's anesthetic management.

To allow ATTENDING to critique a physician's plan intelligently, a flexible appreciation of *risk* on several levels must be incorporated into the system, both to let it assess the relative merits of different approaches to the patient's management, and to let it discuss any associated risks in a natural way.

This paper first gives an example of how the ATTENDING system critiques a plan. Then it describes the heuristic approach to risk analysis which ATTENDING has taken. This approach involves three basic principles, each of which is discussed in detail.

© 1982 by the Electrical and Electronics Engineers, Inc. Reprinted, with permission from the Proceedings of the Sixth Annual Symposium on Computer Applications in Medical Care, Washington, D.C., October 30–November 2, 1982, pp. 697–701.

2. An Example: Critiquing an Anesthetist's Plan

ATTENDING is currently being used in a tutorial mode that allows anesthetist self-assessment. In this mode, the system itself describes a hypothetical patient, after which the anesthetist outlines his plan and receives the system's critique.

This mode allows the system to be useful clinically while its database is still small (since it only need know the anesthetic implications of those medical problems involved in the hypothetical cases). Once the system design is evaluated and finalized, the database will be expanded to allow more general use.

In its current mode, the system first describes a patient, for instance:

A SEVENTY THREE YEAR OLD MAN WITH A HISTORY OF CHRONIC HEPATITIS, MILD RENAL FAILURE, AND THREE PRIOR HEART ATTACKS, COMES TO EMERGENCY SURGERY FOR POSSIBLE ISCHEMIC BOWEL.

After reading this hypothetical case description, the anesthetist outlines his plan. For example, in this case, he might propose:

1. induction of anesthesia with thiopental,
2. rapid sequence intubation using succinylcholine,
3. maintenance of anesthesia with enflurane, curare, nitrous oxide, and oxygen.

The ATTENDING system currently critiques this plan as follows:

THERE ARE A NUMBER OF BASIC PRINCIPLES INVOLVED IN THIS PATIENT'S MANAGEMENT. FIRST, IN THE PRESENCE OF CORONARY ARTERY DISEASE, ONE MUST ATTEMPT TO PREVENT HYPOTENSION AND ISCHEMIA. SECOND, A PATIENT WITH A FULL STOMACH MUST BE PROTECTED FROM ASPIRATION. THIRD, IN A PATIENT WITH CHRONIC RENAL FAILURE, ONE MUST CONSIDER THE PROBLEM OF RENAL TOXICITY AND COMPROMISED RENAL EXCRETION OF DRUGS. ALSO, IN THE PRESENCE OF LIVER DISEASE, ONE MUST ATTEMPT TO PREVENT HEPATIC DAMAGE.

INTUBATION OF THIS PATIENT IS DEFINITELY APPROPRIATE BECAUSE OF A FULL STOMACH. RAPID SEQUENCE INTUBATION IS A REASONABLE APPROACH. THIS MIGHT HOWEVER, HAVE THE RISK OF CARDIAC COMPROMISE. AN ALTERNATIVE APPROACH TO RAPID SEQUENCE INTUBATION WOULD BE INTUBATION AFTER INDUCTION WITH CRICOID PRESSURE. THIS HAS THE ADVANTAGE OF HELPING AVOID HYPOTENSION. INTUBATION AFTER INDUCTION WITH CRICOID PRESSURE MIGHT WELL BE PREFERABLE FOR A PATIENT WITH CORONARY ARTERY DISEASE, IF SEVERE, SINCE IT ALLOWS MORE GRADUAL INDUCTION WITH LESS DANGER OF SUDDEN CARDIAC DECOMPENSATION.

IN REGARD TO INDUCTION, INDUCTION USING THIOPEN-
TAL INVOLVES THE CONCEIVABLE RISK OF HYPOTENSION DUE
TO CARDIAC DEPRESSION. AN ALTERNATIVE APPROACH TO
INDUCTION USING THIOPENTAL WOULD BE AN INHALA-
TIONAL INDUCTION.

FROM THE STANDPOINT OF MAINTENANCE, THE CHOICE OF
ENFLURANE COULD HAVE AT LEAST THE THEORETICAL RISK
OF RENAL TOXICITY DUE TO METABOLIC RELEASE OF FLUO-
RIDE ION. AN ALTERNATIVE WOULD BE HALOTHANE.
HERE, HOWEVER, IN A PATIENT WITH LIVER DISEASE, THERE
IS THE THEORETICAL RISK OF HALOTHANE INDUCED
HEPATOTOXICITY.

LOOKING AT OTHER ASPECTS OF THE PROPOSED PLAN, CUR-
ARE IS WELL SELECTED FOR A PATIENT WITH CHRONIC RENAL
FAILURE. ONE CONSIDERATION WORTH MENTIONING, HOW-
EVER, IS THAT CURARE COULD HAVE THE RISK OF HYPOTEN-
SION DUE TO HISTAMINE RELEASE.

This critique has several features. First, it starts with a short discussion
of the pertinent management principles. Then, in the analysis of the plan
itself, ATTENDING tries to discuss both the *positive* and *negative*
aspects of the plan, thereby articulating any criticisms in as positive a
context as possible.

3. Conventional Risk Analysis

As can be seen, an appreciation of the various risks involved in a patient's
management is at the heart of ATTENDING's analysis. Before discussing
how the ATTENDING system handles risk, however, let us briefly consider a more conventional approach to risk.

The conventional approach to risk analysis, as used in economic cost/
benefit analysis and clinical decision analysis [2], attempts to reduce the
risks involved in decision-making to *numbers* to allow precise comparison. This analysis typically consists of the following steps for each choice
in the decision-making process:

1. defining all possible *outcomes* i,
2. assigning some *value* (V_i) to each outcome,
3. determining a *likelihood* (L_i) for each outcome,
4. computing an "expected value" for that choice, $\Sigma L_i * V_i$.

The difficulties inherent in this general approach are well recognized.
First of all, it is hard to assign precise values to many outcomes, for
instance: loss of life, loss of a limb or an eye, bronchospasm, hypotension,
and other such medical complications of varying severity. Also, the likelihoods of the various risks are not easy to determine precisely. Different
studies often produce quite different figures. Furthermore, it is often hard

to know how a reported complication rate applies to a given patient with a particular set of medical problems.

Granting these uncertainties, it is doubly difficult to take such numbers and add, multiply, and otherwise manipulate them, and still have a good feeling for the exact meaning of the result.

Systems designed using this approach often produce a ranked list of alternatives, with each choice having an associated score indicating how "good" that choice is. The user may be baffled by exactly how to interpret each score, and by what to make of a particular difference between two scores.

In contrast to this conventional, numerical approach to risk analysis, a physician deals with problems of medical management every day with only an approximate feel for the magnitudes of the various risks involved. Instead, he takes a *heuristic approach* to risk analysis, as indeed do people in general in their everyday lives. Certainly it is useful to try to define all the various risk parameters as accurately as possibly, as is attempted, for instance, in clinical decision analysis. At the same time, however, it is equally worthwhile to work on incorporating a more heuristic approach to risk analysis into the machine. The development of the ATTENDING system is a step in this direction.

4. Three Principles for Heuristic Risk Analysis

Instead of reducing the risks of various alternatives to numbers to allow precise comparison, the ATTENDING system bases its approach to risk on the following *three principles:*

1. Rough general criteria are used to eliminate obviously poor choices.
2. Domain-specific knowledge is used to focus attention on the most clinically appropriate alternatives.
3. The relevant risks and benefits are then discussed in a pertinent, natural way.

Figure 1 illustrates schematically the application of these three principles. Here, starting with five possible alternative approaches (A–E), Principle I is first applied to eliminate two obviously poor choices, B and E. Next,

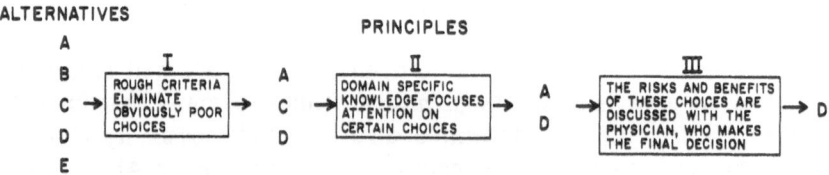

Figure 1. The application of the three principles.

using Principle II, attention is focussed on two of the remaining choices, A and D. The risks and benefits of these are then discussed (Principle III) with the physician, who makes the final choice. This selection process is applied at each level of decision and sub-decision in analyzing a plan.

The remainder of the paper discusses these three principles in turn, and describes specifically how ATTENDING implements each.

PRINCIPLE 1: *Rough Criteria Eliminate Obviously Poor Choices*

If one outlines all possible ways to accomplish a task, there are frequently choices which are clearly inferior. When a person is the decision-maker, many of these poor choices may not even come to mind. They are weeded out at some unconscious level. When a machine is evaluating the alternatives, however, any selection capability of this sort must be explicitly built in.

The use of *rough estimates of the "magnitude" of a risk* allows ATTENDING to achieve this type of "broadbrush" selection capability. For this purpose, each risk is assigned a rough magnitude: LOW, MODERATE, HIGH, or EXTREME. Thus, just as computerized diagnostic systems [3,4] use rough estimates of diagnostic likelihood to drive their analyses, ATTENDING uses rough estimates of risk.

Several concrete examples may help illustrate how these risk magnitudes are assigned:

1) An example of a LOW magnitude anesthetic risk is that of using enflurane, an inhalational anesthetic, when a patient has renal failure (RF). Enflurane releases fluoride ion metabolites, which could theoretically cause renal damage. In practice, however, the quantities released by enflurane are probably insufficient to cause problems. Even so, for a patient with RF, most anesthetists would probably not use enflurane if there were a reasonable alternative, as there usually is.

2) An example of an EXTREME magnitude is the risk that a patient with a full stomach might aspirate (regurgitate and inhale his stomach contents), if his airway is not protected with an endotracheal tube. Aspiration is frequently fatal.

3) An example of a HIGH magnitude is the risk of using succinylcholine in a patient with a pentrating eye wound. Succinylcholine increases intracocular pressure and could thereby lead to loss of the eye.

In addition to risks, management choices may also involve *benefits*. In ATTENDING, benefits too are characterized by a rough magnitude, and are treated internally as "negative risks." In the domain of anesthesia, benefits seem to fall mostly in the LOW to MODERATE range. An example is the advantage (benefit) of inducing anesthesia with ketamine when a patient has asthma, since ketamine is a bronchodilator.

Formally, the "magnitude" of a risk, as used in ATTENDING, corresponds to a rough estimate of its "expected value": its likelihood multiplied by the value of its outcome (i.e., a rough estimate of L_i*V_i). *Clinically*, these risk magnitudes seem to conform naturally to the way an anesthetist thinks about risks, and are therefore quite easy to assign.

When manipulating risk magnitudes internally, ATTENDING makes certain simplifications. If a choice incurs several risks, the risk magnitudes are *not* added together to determine the total risk magnitude of that choice. Rather, the largest risk magnitude is used. Thus if a given choice involves one LOW risk and two MODERATE risks, the total risk magnitude assigned to that choice is MODERATE.

Therefore, even though there might be several risks and sub-risks involved in a choice being evaluated, that choice is always characterized by a single risk magnitude. As a result of this simplification, rough "broadbrush" comparison between two choices is very straightforward. If the risk magnitudes are not the same, then one choice is clearly inferior and can be discarded. The only exception is that the physician's choice is retained, even if it is inferior, since it will serve as a basis for the later critique.

Thus, for example, if four choices exist, two with LOW risk, one with MODERATE risk, and one with HIGH risk, the MODERATE and the HIGH risk choices can be discarded, leaving only the two low risk choices for further analysis. For instance, in paragraph 2 of the example critique, the possibility of not intubating the patient (which has the EXTREME risk of aspiration) is discarded early in ATTENDING's analysis and never mentioned in the critique.

PRINCIPLE 2: *Domain-Specific Knowledge Focuses Attention on the Most Clinically Appropriate Alternatives*

Using rough estimates of risk as discussed above, ATTENDING is able to discard obviously poor choices, and is left with alternative choices which have risks of equal "rough magnitude." This broadbrush discrimination, however, may not be sufficient to let the system focus on the most clinically appropriate alternatives for discussion with the physician. More specific knowledge may have to be brought to bear.

To allow this more selective focusing of ATTENDING's attention on the clinically most appropriate alternatives, the ATTENDING system uses *"Contextual Preference Rules"* (CP rules). CP rules are designed to let the system look at approaches which have the same risk magnitude (LOW, MODERATE, etc.) and selectively focus on a preferred approach in a given context. Each CP rule has *four* parts, which state that 1) in the context of a particular medical problem, 2) a certain specified technique is most likely preferable to 3) another specified technique. The CP rule also states 4) the reason for the preference.

For instance, in the current ATTENDING implementation, there is a CP rule stating that 1) in the presence of coronary artery disease, 2) intubation after induction with cricoid pressure may be preferable to 3) rapid sequence intubation because 4) in allows more gradual induction with less danger of sudden cardiac decompensation. (In paragraph 2 of the example critique, one can observe this CP rule being used in discussing the plan.) As can be seen, a major advantage of using CP rules is that not only can the system focus on a preferred approach, it can also *justify its choice* since it knows the reason for the preference.

Thus, CP rules play two roles. First, they allow ATTENDING to focus selectively on certain approaches. Second, if the physician has proposed a "less preferable" choice, a CP rule gives ATTENDING a reason it can use in justifying its suggestion of the "more preferable" approach.

To understand the role played by the CP rules, it is instructive to discuss how their use differs from trying to achieve similar selective focusing by "fudging" (i.e., further refining) the risk magnitudes. In other words, why have CP rules at all? Why not, for instance, in the presence of coronary artery disease, somehow adjust the risk magnitudes associated with "rapid sequence intubation" and "intubation after induction with cricoid pressure" so that they are no longer equal, and so that the latter is preferable, purely from the standpoint of risk magnitudes?

There are several disadvantages to such "fudging". First of all, once one starts systematically "fudging" risk magnitudes in this way, even the system designer can rapidly lose sight of the various permutations of exactly when and where one technique is preferable to another, and why. Equally important, by burying the preference information in these weighting factors, this information is irretrievably lost to the system itself. *It can no longer explain and justify its preferences to its physician user.*

Of course, one might argue that the use of CP rules has the *practical effect* of "fudging" the risk magnitudes, since the CP rules are superimposed upon the risk magnitude analysis, and serve to refine it. Perhaps so, but by using CP rules, this is done cleanly. The risk magnitude analysis is kept simple and is easily understood. The CP rules themselves state explicitly where and when they operate. And most important, as mentioned previously, they allow the system to explain and justify the preferences.

PRINCIPLE 3: *Discussion of Relevant Risks and Benefits in a Pertinent, Natural Way*

The ability to discuss the relevant risks and benefits in a way that seems natural to the physician is of paramount importance. If the system cannot do this, then its impact will be greatly undermined. We have already shown how CP rules allow ATTENDING to explain certain of its recommendations to the physician. But even more than this is needed. In particular, *the system must tailor its discussion of each risk to the physician's own perception of that risk.*

Risks have many facets. Not only do they have different magnitudes, they also have many other characteristic features. These features influence how the physician thinks about each risk, and therefore how that risk is best described. For instance, some risks are virtually certain, and others are remote. Some are unanimously recognized, others are controversial. Still others may be at best theoretical, and of little practical importance. Some may be so obvious as to be implicit in any discussion involving them.

These features might most aptly be called "pragmatic features" of risks. Each risk has its own set of "pragmatic features". ATTENDING must know about these pragmatic features if it is to discuss the risks intelligently with the physician.

Among the pragmatic features used by ATTENDING to characterize risks are:

1) *IMPLICIT:* Some risks are implicit in the techniques being described. An example is the risk of aspiration while doing a mask induction with circoid pressure for a patient with a full stomach. Another example is the cost and risk implicit in preoperatively performing a tracheostomy to secure a patient's airway. ATTENDING must know about these risks to be able to evaluate alternative approaches to intubation. The system would sound naive, however, if it discussed such implicit risks each time it mentioned the technique involved. Thus, risks with the IMPLICIT feature are weighed internally during ATTENDING's analysis, but are not mentioned to the physician in the discussion of his plan.

2) *THEORETICAL:* Some risks are really only of theoretical importance in that they probably never cause major harm, even though they may often influence a patient's management. An example is the risk, discussed earlier, of using enflurane in a patient with renal failure. Renal damage by enflurane has never been documented. When these risks are discussed by ATTENDING, they are couched in a phrase like, "there is *at least the theoretical risk of* —." This phrasing tailors the discussion to the physician's own perception of that risk.

3) *REMOTE:* Some risks, although acknowledged as real, may be perceived as remote possibilities. An example is the risk of provoking bronchospasm by inducing anesthesia with thiopental. It seldom actually happens. ATTENDING's discussion of such a risk is couched in a phrase like, "there is *the conceivable risk of* —."

Pragmatic features such as these are necessary because the physician has different mental models of different risks. To discuss these risks with him in a natural way, wording must be chosen carefully to tailor the discussion to his mental models. If these pragmatic features are ignored, and ATTENDING discusses all risks in the same terms, then the system sounds like a school child reciting facts that it has memorized but doesn't fully understand.

If, on the other hand, the pragmatic features are used, and the discussion of each risk is appropriately modified or qualified, the anesthetist

Principles	Implementation
Rough Criteria Eliminate Obviously Poor Choices	rough estimates of the "magnitude" of a risk
Domain-Specific Knowledge Focuses Attention on the Most Appropriate Alternatives	contextual perference rules
Discussion of Relevant Risks and Benefits in a Natural Way	pragmatic features

Figure 2. The three basic principles for heuristic risk analysis and ATTENDING's current implementation of each.

usually doesn't even notice that the discussion of different risks is being phrased differently.

5. Summary

In summary, the paper has described a heuristic approach to risk analysis in medical management. Three underlying principles have been outlined, and the way in which the ATTENDING system implements each of these principles has been described. The relationship between each underlying principle and its implementation in ATTENDING is illustrated schematically in Figure 2.

It is important to recognize that there is a distinction between the three principles, per se, and their current implementation in the ATTENDING system. As the development of ATTENDING continues, the *implementation* of the principles may well be refined and augmented. It is anticipated, however, that the underlying principles themselves will remain constant.

References

[1] Miller PL: ATTENDING: A system which critiques an anesthetic management plan. Proc. Am. Med. Informatics Assoc. Congress 82, San Francisco, May 1982.
[2] Weinstein MC, et al.: Clinical Decision Analysis. Philadelphia: W. B. Saunders, 1980.
[3] Kulikowski C: Artificial intelligence methods and systems for medical consultation. IEEE Transactions on Pattern Analysis and Machine Intelligence. 1980, Vo., PAMI-2, No. 5.
[4] Shortliffe EH, Buchanan BG, Feigenbaum EA: Knowledge engineering for medical decision making. A review of computer-based clinical decision aids. Proc. IEEE 1979; 67:1207–1224.

V. COGNITIVE MODELS

20

Clinical Problem Solving:
A Behavioral Analysis
Jerome P. Kassirer and G. Anthony Gorry

To extend the understanding of the clinical problem-solving process, we
have analyzed the tape-recorded behavior of experienced clinicians engaged
in "taking the history of the present illness" from a simulated patient. We
showed that specific diagnostic hypotheses were generated often with little
more information than presenting complaints, that testing of diagnostic
hypotheses consisted of various case-building strategies for corroborating
and discrediting hypotheses, and that the process of information gathering
included techniques to evaluate the validity of data and assess the need for
immediate action. Overall strategies were more difficult to discern but
included a focused approach, a systemic exploration method, and a chronologic technique. The data have potential value in medical education and
in developing computer programs to simulate the diagnostic process.

Problem solving is a critical function performed by the physician, yet the
processes that underlie this intellectual achievement are poorly understood. Attempts to explicate the components of the diagnostic process,
one of the key elements of clinical problem solving, have been derived
chiefly from the reflection of experienced clinicians on their own problem-solving processes [1–10]. Such reflections have yielded many important principles and a general framework within which students can
develop a comprehension of the decision-making tasks of the profession.
However, because these principles are general and because so many situations arise with so many different possible approaches, the application
of the principles is anything but straightforward and is rarely specified.
This imprecision fosters the belief that the problem-solving processes of
clinicians cannot be scrutinized, and, because of the mystery surrounding
the concept of clinical judgment, the seasoned clinician relies on the student's ability to deduce from observation what he as an experienced physician practices but is unable to articulate.

Reprinted with permission from *Annals of Internal Medicine, 89,* pp. 245–255. Copyright
1978 by the American College of Physicians.

Investigating the diagnostic process is difficult, particularly because of the wide scope encompassed by the subject material and the uncertainties inherent in medical information. Nonetheless, studies of problem solving in domains as complex as logic [11], chess [12], and cryptarithmetic [13] have disclosed important principles of human problem-solving behavior. Despite the complexities of the medical area, we anticipated that the application of methods in medicine that are comparable to those used by behavioral psychologists in these other domains might provide useful data. Because "taking the history of the present illness" requires a flexible, unstructured approach, we investigated this aspect of clinical problem solving. We used a method of study that combines protocol analysis (the recording and study of diagnostic problem-solving behavior) with introspection to obtain an analysis of the strategies that clinicians use. The subjects were experienced clinicians accustomed to dealing with difficult diagnostic problems and were selected on the assumption that they are experts in clinical problem solving. Detailed examination of the behavior of these clinicians has yielded important principles underlying the process of unstructured history taking.

The benefits of a deeper, more explicit understanding of this process are undeniable. An accurate delineation of principles and strategies should improve the process of developing clinical reasoning in young physicians. Such principles and strategies could also be used in the development of better measures of the quality of medical decision making and could, therefore, improve medical care. Finally, advances in computer-aided clinical decision making depend on the development of better understanding of human problem solving in the clinical setting.

Experimental Approach

Goal of Study

The study was designed to identify some of the specific tactics and general strategies constituting the problem-solving behavior of experienced clinicians as they were engaged in "taking the history of the present illness." Many elements other than those associated with diagnostic problem solving contribute to the interaction between patient and physician during the history-taking process. Factors such as effective communication between doctor and patient, the psychosocial importance of the interaction, and the physician's evaluation of the accuracy of individual symptoms reported by the patient are all important, but these aspects were not assessed in this study.

Selection of Experimental Subjects

The six clinicians studied were members of the Department of Medicine at the New England Medical Center Hospital. The group consisted of four nephrologists, a cardiologist, and a gastroenterologist.

Obtaining the Protocols

One of the authors played the role of the patient. This individual, here called the *respondent,* used as the basic "script" the record of a patient who had been hospitalized, and he answered the questions of the clinician taking the history of the present illness. Each subject was given the following instructions.

1. You will be asked to "take the history of the present illness" from a simulated patient. Initially we will give you the patient's age, sex, and chief complaints.
2. The conditions of the experiment are as follows.
   a. The respondent will not directly simulate the patient in giving answers to your questions. Instead, he will answer any questions you ask about the patient from his knowledge of the patient's history and clinical course.
   b. Your questions should be as specific as possible. General questions such as "What problems has the patient had in the past?" will not be answered.
   c. When you ask a specific question, please provide the reason for asking it.
   d. You are requested to tell what you learned from the answer to each question and to report any hypotheses you are considering.
   e. When you believe you have reached a final diagnosis or have gone as far as you can, please summarize your diagnostic impressions.

The clinicians had no difficulty following these instructions, but from time to time it was necessary to remind them to explain the reasons for their questions. Questions were answered until a conclusion was reached or until the information from the record was exhausted. As a first approximation, we assume that each subject took the history of the present illness of the same patient, although some obtained information that others missed and vice versa.

Selection of Case Material

The patient selected for study was well know to one of the authors, thus enabling him, with frequent reference to the record, to provide pertinent information in his role as the "respondent." The case was selected to conform to the following criteria.

1. It was one that a physician might encounter in the everyday course of clinical practice, not a "trick" case involving an obscure disease or a rare manifestation of a common disease.
2. The clinical presentation and history were both suggestive of and consistent with the final diagnosis. No effort was made, however, to find a "classic" case for a particular disease. Both aspects were important

because the experiment was designed as a means to capture data about "standard" methods of clinical cognition, not to test diagnostic skill.
3. The case was rich in history with enough data available about the medical history to provoke consideration of a number of possible diagnostic options.

A summary of the patient's findings is provided here to illustrate how this case fits the criteria, but the clinician subjects did not have access to this summary. Data were given to the subjects only when they requested specific information.

Patient Summary

The patient was a 57-year-old nurse admitted to the New England Medical Center Hospital in 1967 complaining of nausea, vomiting, abdominal pain, and frequency of urination. At the age of 40 she had a left nephrolithotomy, hysterectomy, and cholecystectomy; urethral catheters had been used after all three procedures. In the 15 years before admission, she had taken approximately six Empirin® tablets per day for chronic headaches. Eight months before admission to the New England Medical Center Hospital she was admitted to another hospital with complaints similar to those experienced before the current admission. At that hospital she was found to have pyuria, hematuria, and 4+ proteinuria, and the nonprotein nitrogen concentration was 51mg/dl. Two intravenous pyelograms were done, but neither visualized; a retrograde pyelogram was said to show contracted kidneys bilaterally. She was given 3 units of blood. Urine cultures were positive on two occasions for enterococci in colony counts greater than 100,000 organisms per millilitre; treatment with ampicillin was instituted, and the patient was discharged. The patient's symptoms improved, but transfusions were given monthly.

Four weeks before admission, nausea and vomiting recurred. Three weeks before admission, she developed streptococcal pharyngitis and was treated with penicillin. In the 2 weeks before admission she lost 4.5 kg, developed periorbital adema and shaking chills, and complained of frequency and urgency of urination.

When the patient was admitted to the New England Medical Center Hospital the vital signs were normal, and skin turgor was fair. Except for left costovertebral angle tenderness, results of the physical examination were unremarkable. Laboratory studies showed that she was anemic (hematocrit reading, 20%), azotemic (serum creatinine, 9.4 mg/dl; blood urea nitrogen, 96 mg/dl), and acidotic (arterial pH, 7.1; HCO_3, 5 meq/litre; Pco_2, 15 mm Hg). Urinalysis showed 1+ protein, and the sediment was filled with leukocytes. Urine culture contained enterococci in colony counts greater than 100,000 organisms per millilitre. Intravenous pyelogram showed poor visualization and small kidneys bilaterally. The patient was treated with intravenous fluid and antibiotics; after 3 weeks in the hospital all symptoms had cleared, serum creatinine had fallen to 6.0 mg/dl, the urine had cleared of leukocytes, and bacteriuria had disappeared. Diagnoses made were chronic renal disease—either pyelonephritis, analgesic nephropathy, or both; acute pyelonephritis superimposed on chronic renal disease; chronic renal failure; metabolic acidosis; and anemia secondary to chronic renal disease.

Analysis of Protocols

Detailed analysis of protocols of the participating clinicians shows that the clinicians tried various hypotheses as explanations of the patient's problem—hypotheses that "fit" the case at hand more or less well. Piece by piece, they assembled the evidence for and against competing hypotheses until one hypothesis seemed clearly better. When findings seemed inconsistent with a hypothesis, they searched for an explanation for the inconsistency or they rejected the hypothesis in question.

We shall examine the protocols in terms of major elements of the clinicians' problem-solving process: hypothesis activation, hypothesis evaluation, and information gathering. This separation is somewhat artificial, of course, because hypothesis activation, hypothesis evaluation, and information acquisition interact repeatedly in the history-taking process.

Hypothesis Activation

The most striking aspect of the history-taking process revealed by the protocols is the sharp focus of the clinicians' problem-solving behavior. The subjects generated one or more working hypotheses early in the history-taking process when relatively few facts were known about the patient. At a time when the clinician was aware only of the age, sex, and presenting complaints of the patient, he often immediately introduced a hypothesis, a phenomenon that suggests that the pattern of age, sex, and these particular presenting complaints had activated this hypothesis (Figure 1). Hypotheses were most often activated by introduction of individual symptoms by the respondent but were also triggered by pairs or combinations of symptoms or by mention of a cause or a complication of a disease. In some instances hypotheses were activated when there was a notable discrepancy between new information obtained from the respondent and an already existing hypothesis. Hypotheses were also generated through a "stringing together" of concepts: In one instance, for example, analgesic abuse was hypothesized in response to the mention of papillary necrosis, which in turn had been hypothesized in response to the mention of diabetes mellitus, which in turn had been hypothesized in response to a possible diagnosis of pyelonephritis.

A wide spectrum of hypotheses was generated from clinical findings. On one end of the spectrum, provision of new information activated no new hypotheses when the clinical finding was quite general, for example, a symptom such as nausea. When the finding was somewhat more spe-

HYPOTHESIS GENERATION

RESPONDENT: This is a 57-year-old woman who is admitted to the hospital with the chief complaint of nausea, vomiting, abdominal pain and frequency of urination.

CLINICIAN: First I'm going to ask some questions about the character of her urinary stream because I'm thinking in terms of *infection in her lower urinary tract*. Did this patient notice any blood in her urine?

RESPONDENT: No, she didn't.

CLINICIAN: That she didn't have gross hematuria makes me turn away from one possibility—that she might have *passed a stone in association with infection*. She might have had a *hemorrhagic cystitis* but that makes it unlikely, just at first cut. You said she had frequency—did she have pain on urination? I'm asking that in terms of also *inflammation of the bladder*.

RESPONDENT: She did complain of some burning on urination.

CLINICIAN: Now again continuing along the infection line, I'm going to ask whether she had a fever just in terms of *generalized infection*.

Figure 1. The rapidity with which experienced clinicians generate hypotheses concerning the patient is illustrated by this excerpt from a protocol. Hypotheses considered by the clinician are italicized. The specificity of these hypotheses is noteworthy in view of the small number of facts available about the patient. All of these excerpts have been preserved in their original conversational style.

cific, however, such as fever and chills, a general hypothesis such as "infection" was typically activated. At the other end of the spectrum, a clinical finding closely correlated with a specific disease entity activated a specific hypothesis (such as dysuria activating the hypothesis of urinary tract infection). Finally, it was observed that some diagnostic hypotheses, although deemed unlikely by the clinicians, were considered early and earnestly. On analysis it was apparent that these hypotheses, such as sepsis, urinary tract obstruction, and salt depletion, were entertained solely because of the potential benefits of therapeutic intervention. This aspect of the process, the therapeutic implications of unlikely diagnoses, is discussed in further detail below.

Another striking aspect of the process observed was the small number of active hypotheses maintained by all the clincians. Although the total number of hypotheses entertained throughout the diagnostic session varied from 14 to 37 (Table 1), the physician participants indicated that they maintained only four to 11 hypotheses active at any one time.

The process of hypothesis activation dominated the early part of the diagnostic session as the physicians searched for some explanation of the findings and for a context in which to proceed. Later in the session the emphasis was on hypothesis evaluation rather than hypothesis activation. Identifying a context through the process of hypothesis activation appears to be one of the critical features of the diagnostic process. When a hypothesis was activated, the physician appeared to use this context— that is, his concept of a disease, a state, or a complication as a model with which to evaluate new data from the patient. Such a model provides a basis for expectation; it identifies the relevant clinical features that should prove fruitful for further investigation. Clearly, the ability of the physician to carry out these cognitive tasks is a function not only of his grasp of the problem-solving methods but also the extent of his knowledge of disease entities. The relative importance of the knowledge base, the problem-solving skills, and the interaction of these two entities in the diagnostic process is not known and was not subjected to analysis here.

Hypothesis Evaluation

General Approach. All the subjects used a common approach to evaluating hypotheses previously activated. Requesting and assessing new information, they rejected some of the initial hypotheses, substituted specific hypotheses for more general ones, and selected a few specific hypotheses for detailed and critical testing. They refined these specific hypotheses by a process we have referred to as case building. After selecting one or two hypotheses as candidates for a final diagnosis, they

Table 1. Summary Analysis of Six Protocols

Clinician	1	2	3	4	5	6
Specialty	Nephrology	Nephrology	Nephrology	Nephrology	Gastroenterology	Cardiology
	←			no.		→
Object of questions						
Finding	21	11	18	29	43	61
Temporal relationships	8	14	16	7	12	7
Organ system	8	6	12	27	43	87
Severity	9	5	11	3	5	15
Complication	11	7	9	4	3	6
Predisposition	7	1	2	5	9	8
Need for action	7	5	5	0	5	5
Total questions asked	35	25	36	42*	80	117
Number of question at which correct diagnosis was first mentioned	9	5	8	11	19	22
Number of question at which firm diagnosis was made	19	19	19	36	52	70
Total hypotheses considered	23	21	20	14	26	37
Maximum nuber of hypotheses active at any one time	6	6	4	5	11	9

*Short gap in tape. Approximately four to six questions lost.

assessed the adequacy of the hypotheses as to whether they encompassed all the clinical findings and provided a coherent explanation of predisposing or causative factors and complications. These strategies are considered below.

The Process of Case Building. Case building consisted of tactics used to evaluate and refine hypotheses, incorporate new data into existing hypotheses, and modify or eliminate hypotheses. An example of case building in which an elementary hypothesis is successfully replaced with a more specific version is illustrated in Figure 2. As noted earlier, the physicians appeared to evaluate those hypotheses against their model of a disease entity (such as a disease, a syndrome, or a complication). In doing so, they matched the pattern of findings in the patient to their concepts of various features of diseases including the known clinical characteristics, the evolutionary changes in clinical features, the predisposing factors, and the complications. The evaluation of a hypothesis as an explanation of the clinical findings in a patient is a complex problem requiring the assessment of how many findings caused by a given disease are present and how many expected findings of a given disease are absent. The problem is particularly difficult because the evidence available is characterized by various degrees of reliability.

REFINEMENT OF A DIAGNOSTIC HYPOTHESIS

CLINICIAN: So she had something 17 years ago and then 8 months ago . . . I'd like to hear a little more about the symptoms that she had, the onset and the duration. I'd like to go back now; I'm interested now not in the most recent episode which is what brings her here. My approach would be to go back to the beginning of the story because I don't have a proper focus on it by just feeding on this episode because the fact that she's here now isn't as interesting to me as to get the evolution of what could be a *single recurrent illness.* I'm about to test that hypothesis by going back. So what was her initial illness like 17 years ago?

RESPONDENT: She had a kidney stone on the left side 17 years ago which was removed surgically. That's the only information we have about that particular episode.

CLINICIAN: Yes, I mean I would probe; I would like to know everything I can about each of these three episodes. I can tell you what my thinking now is. The likelihood that this is now *urinary tract infection* has gone up considerably, but not as much as if that episode were 2 or 3 years ago. That is, the 17-year lag is compatible with long-standing . . . either recurrent infection which was asymptomatic or at least a predilection to the *recurrent infection* due to an anatomical abnormality, but 17 years is a long time and it's pushing me.

Figure 2. The way in which an experienced clinician refines a diagnostic hypothesis, narrowing it as more facts become available, is illustrated in this excerpt.

Ideas concerning the way in which the case-building process takes place have been discussed elsewhere. Techniques including Bayes rule [14,15], Boolean algebra [3], template matching [16], and others [17–20] have been used as models of this process. Some of the heuristic methods used by physicians were, however, readily identified in the protocols. A *confirmation* strategy was that with which the clinician tried to prove a hypothesis by matching characteristics of the disease or clinical state under consideration. Although in some instances hypotheses could be confirmed through the results of laboratory tests alone, in the majority of instances a hypothesis was substantiated through the accumulation of many pieces of evidence that proved compelling only when seen together. Accordingly, much data had to be gathered about many aspects of the patient's condition before the physicians were convinced that the correct diagnosis had been made. Some segments in the protocols contained as many as seventeen questions asked by the physician to confirm a hypothesis. Another example of case building in which this strategy is used is shown in Figure 3.

Just as there are many ways to confirm a hypothesis, there are many ways to eliminate one. An *elimination* strategy used questions about findings that are so often found with a given disease that their absence weighs heavily against the hypothesized disease. In many instances, the elimination strategy was used to remove from further consideration a hypothesis that was unlikely but that could not be ignored. Several pieces of historical information rather than a laboratory result often serve this purpose. An example of this technique is shown in Figure 4. A *discrimination* strategy was a special kind of elimination strategy, but its purpose was to distinguish one hypothesis from another—for example, to distinguish an acute disorder from a chronic one. There appeared to be relatively few instances in which a clear-cut discrimination was possible on the basis of raw clinical data, and most often the physician had to discriminate between two competing hypotheses indirectly. An example of the discimination between chronic pyelonephritis and chronic glomerulonephritis is shown in Figure 5.

An *exploration* strategy was used to refine a hypothesis both by making it more specific and by checking for complications, related conditions, or unrelated disorders when an accepted hypothesis was thought to explain the known findings. After the physician had refined a broad hypothesis such as "chronic renal disease" or "acute abdominal problem," for example, to chronic pyelonephritis or acute pancreatitis, he often reverted to this general exploration strategy. The exploration strategy was sometimes in the form of a general "review of systems," but more frequently it was also directed by the physician's expectations of findings based on the hypotheses he had already formulated. This strategy was frequently used in the search for complications of the disease being considered that had not yet been uncovered in the history-taking process. All the physicians

BUILDING A CASE FOR ANALGESIC NEPHROPATHY

CLINICIAN: Had she been subject to headache?

RESPONDENT: She had a long history of headaches.

CLINICIAN: Was she treated?

RESPONDENT: Yes, she has taken some medication for some time.

CLINICIAN: Do we know the nature of this medication?

RESPONDENT: She took Empirin tablets.

CLINICIAN: How many did she take?

RESPONDENT: As far as we can tell she took around six tablets per day.

CLINICIAN: Over what period of time?

RESPONDENT: For about 15 years.

CLINICIAN: What I'm concerned about now is whether she has had an abuse of this compound. (Interposed here are several questions concerning a history of hypertension, visual disturbances, seizures, psychiatric disorders, and respiratory symptoms.) Obviously, the most cogent thing we've picked up in this review is the very heavy abuse of Empirin which I think could well be related to the problem she had with kidney stone; could well, in fact, have been a sloughed papilla from papillary necrosis. The episode she's having now could well represent episodes of renal infection or papillary necrosis. (Interposed here are several questions concerning cardiac symptoms, gastrointestinal complaints, menstrual and child-bearing history, joint symptoms, and history of use of other medications.) Just to get some information on her as a person, I'd like to know what her social situation is.

RESPONDENT: She was born in Massachusetts, was a nurse, and runs a nursing home now. She previously smoked, but hasn't recently. She doesn't drink alcohol and drinks occasional tea and coffee.

CLINICIAN: The fact that she's a nurse again sort of heightens my suspicion that she may, in fact, have been abusing drugs for some period of time. Because of the access to such agents, it is assumed that the incidence is higher.

Figure 3. The use of evidence that suggests, but does not directly confirm, a hypothesis is illustrated here. Using a "case-building" strategy, the clinician gathers information that in total strongly suggests the diagnosis of analgesic nephropathy.

STRATEGY OF ELIMINATION

CLINICIAN: One other thing I might be interested in in this lady with what you told me is the fact that she continued to have white cells in her urine. Although it's completely consistent with chronic pyelonephritis I would be interested in getting a TB culture on her just because of her history a long time ago. I would expect them to be negative.

RESPONDENT: TB cultures were performed and were negative.

Figure 4. Often a clinician finds it advantageous to try to eliminate from further consideration hypotheses that cannot yet be ignored. Here is an example of the elimination of the hypothesis of renal tuberculosis (TB) from active consideration.

CLINICIAN: Did she have any history of high blood pressure? The reason I'm asking that question is that in association with certain kinds of renal insufficiency, hypertension is a very common feature.
RESPONDENT: No, she didn't.
CLINICIAN: The answer to that question leads me away from something like chronic glomerulonephritis as causing her renal insufficiency. It's consistent with chronic pyelonephritis. You could have hypertension or no hypertension.

Figure 5. When two diseases or conditions are apt to present with similar findings, the clinician may develop a specialized set of questions with which to distinguish the two. Here the clinician uses the absence of hypertension as an argument for chronic pyelonephritis and against glomerulonephritis.

studied, for example, discovered that the patient had anemia of renal failure, and five of six discovered the complicating metabolic acidosis. In all instances the physicians explored the case well beyond the point at which the principal diagnosis had been established. This process is critical because, as noted before, some of the complications are serious and require urgent therapeutic intervention.

Assembling a Coherent and Adequate Diagnosis. In parallel with the processes by which the physicians built a case toward a final diagnosis, they assessed each diagnosis for *coherence* and *adequacy.* A diagnosis was coherent when physiologic links were appropriate; for example, when the physical finding of poor skin turgor was associated with a history of fluid loss through vomiting. A diagnosis was also considered coherent if predispositions or complications associated with the disease entity were appropriate. Thus, a diagnosis of chronic pyelonephritis was felt to be coherent in the patient discussed here because of the association between chronic pyelonephritis with recurrent urinary tract infection and excessive intake of analgesic compounds.

A diagnosis was considered *adequate* when it encompassed all elementary hypotheses and accounted for all the abnormal (and normal) findings. The physicians strove to attain parsimonious explanations for the findings and to accept a single explanation rather than make two or more diagnoses unless they were forced to do so. Techniques for dealing with information that appeared to be discrepant with the principal hypothesis are discussed below.

Information Gathering

To test the various hypotheses that came to mind during the course of a history-taking session, the clinicians gathered a large amount of information concerning the patient including laboratory tests, physical findings, current and past symptoms and treatment, and various aspects of the patient's lifestyle. The clinicians approached the problem of information gathering in different ways, and their protocols show corresponding differences in the questions they asked and the sequence in which they asked them. Despite these differences, there is a striking similarity among the protocols in the extent to which the clinicians made reference to features of the diagnostic hypotheses other than raw clinical data such as characteristic signs, symptoms, and laboratory findings. These features, listed in Table 1, were explored in depth. *Temporal relations* included an assessment of whether the disorder was acute or chronic, whether it had an episodic character, and what the pattern of disease was over time. Questions directed at an *organ system* consisted of signs, symptoms, and laboratory findings highly correlated with a diseased organ system (such as dysuria or pyuria with urinary infection). Questions concerning *severity* were those directed at determining the overall severity of the patient's illness or assessing the severity of disturbance in function of an individual organ. The questions concerning *predisposition* or *complications* were directed at uncovering either physiologic links (direct cause-and-effect relations) or associations in which a valid correlation exists between a clinical disorder and a cause or complication. The questions concerning a *need for action* relate to therapeutic decision making and are discussed later. Approximately 60–80% of the questions in each protocol were directed at these specific features. This observation suggests that much of the questioning is "hypothesis-driven" in contrast to the general information-gathering approach typical of the "review of systems." The remaining questions were directed at obtaining general information usually considered characteristic of the system review. Data on the number of questions directed at each feature, the total number of questions asked by each clinician, and the hypotheses entertained are shown in Table 1.

There are presumably many reasons why a clinician may ask about one of these features of a particular hypothesis. This analysis is not designed to ascertain why one feature is investigated in a particular circumstance and another is not. It is important to recognize, however, that this short list of features of a disease, physiologic condition, or syndrome seems to be a useful set of concepts around which to organize knowledge concerning a given diagnostic entity. Regardless of the information sought, the clinician appears to seek data related to one of these features.

The Style of Information Gathering. Clearly, the overall style of information acquisition is a complex process that can be analyzed only superficially with the methods used here. Nonetheless, several aspects of this process are apparent. First, the expert nephrologists used a highly directed information-gathering method. Recognizing quickly that the patient had a renal problem, these physicians focused on obtaining pertinent information regarding the kidneys and urinary tract, usually using an extended series of questions to assess the presence of manifestations of renal disease. In addition, as noted in Table 1, the nephrologists asked fewer questions, mentioned the correct diagnosis earlier, made a firm diagnosis earlier, and maintained a smaller number of active hypotheses than the two physicians who were not expert in the patient's illness. Presumably, both the gastroenterologist and the cardiologist would have used a similar directed approach if the patient's condition had lain within their respective fields of clinical expertise. As it stands, however, the behavior of these two experienced clinicians with a clinical problem outside their area of expertise is instructive. Both clinicians asked most of the important questions asked by the nephrologists, but the pattern of questioning was not as directed. These physicians tended to explore more symptoms and findings unrelated to the urinary tract even when the questioning was producing little valuable data. In other words, the cardiologist and the gastroenterologist often reverted to a general review of systems when they had no clear direction in which to proceed with their questions. Both physicians discovered the important history of analgesic abuse: In both instances this finding was uncovered in a general system review directed at headaches or chronic ingestion of medications rather than—as in the case of the nephrologists—through a question directed at determining factors that might have predisposed the patient to unexplained chronic renal disease.

The Expert Witness Concept. Clinicians are often forced to assimilate data about the patient that are largely historical. Except for his own physical examination of the patient and a few current laboratory tests, the physician must work with facts presented by the patient and with reports by other physicians, from other hospitals, or from both sources, of previous symptoms and signs, laboratory studies, diagnoses, and treatments. Because the time course of the illness and possible predisposing factors or previous instances of the disease are important in understanding the current situation, the clinician must attempt to construct an undistorted view of the patient's history in face of potentially unreliable and sometimes incomplete data.

The protocols show that the clinicians made extensive use of one interesting strategy in assessing the validity of data in the history. They frequently sought to identify an *expert witness* whose report of a particular circumstance, procedure, or finding could be accepted as accurate or whose actions could be deemed consistent with a particular interpretation of an event, condition, or circumstance. The technique was used not only to confirm diagnostic impressions but also to assess the severity of previous illness. Examples of such a technique to test the general severity

---

EXAMPLE OF "EXPERT WITNESS" TECHNIQUE

CLINICIAN: Was she hospitalized on that occasion?
RESPONDENT: Yes, she was. Why did you want to know that?
CLINICIAN: Because I would assume to some extent the severity of her symptoms
               might be reflected in whether or not she was hospitalized. It would give
               me an indication of how seriously ill she was at the time.

Figure 6. Because so much of the information about the development of the patient's illness is learned from the patient, the physician uses methods to assess the accuracy and completeness of that information. If it can be established that a competent observer was present at the time in question, reports of that observer's stated impressions or actions can be used to validate clinical data. This general approach has been termed the "expert witness" technique.

of illness are shown in Figures 6 and 7. In the example shown in Figure 7, the clinician assumed that if protein restriction had been prescribed by a previous physician, the patient's renal disease was probably advanced sufficiently to lead to uremic symptoms and that the restricted-protein diet had been used to treat these symptoms.

The availability of an expert witness is, however, only one part of this technique; the quality of the witness looms as critically important in the evaluation of previous data. The clinicians often asked specific questions about the expert witness such as "Who took care of the patient?" "What hospital was she in?" "What diagnosis did they make?" and "What treatment was given?" Often they would evaluate the expert witness on the basis of the action taken, as in the excerpt from a protocol shown in Figure 8.

Purposes of Information Gathering. One purpose underlying the selection of a particular question or set of questions was to construct a plau-

---

EXAMPLE OF "EXPERT WITNESS" TECHNIQUE

CLINICIAN: Was there any modification of her diet?
RESPONDENT: Not that we're aware of.
CLINICIAN: Was there any reduction in the amount of protein in her diet?
RESPONDENT: We're not aware of any change in diet.

Figure 7. See caption to Figure 6.

EXAMPLE OF "EXPERT WITNESS" TECHNIQUE

CLINICIAN: I don't understand her anemia because she's not azotemic enough. She could have an independent cause of anemia. We don't know how severe her anemia was. It was presumably pretty severe if they gave her three transfusions. I mean, people don't lightly do that and the question is if you want to be parsimonious, the anemia logically should go with renal failure. Did they transfuse her inappropriately? That seems unlikely; it's pretty remote that she wasn't fairly anemic.

Figure 8. See caption to Figure 6.

sible hypothesis or hypotheses to account for the patient's presenting signs and symptoms as well as the known laboratory findings. Another purpose of a more immediate concern was to *assess the need for immediate diagnostic* or *therapeutic action.* Questions often focused on the intensity, frequency, and duration of certain signs and symptoms. An example of questions with this purpose is shown in Figure 9. In some instances, the clinician also asked certain questions in order to lay the groundwork for future decisions concerning therapy. The information gained may be of little use in refining the diagnostic hypothesis but may be very valuable in anticipating risks and benefits of certain procedures that may be needed in the near future.

INFORMATION ACCUMULATION FOR DEFINING A NEED FOR ACTION

CLINICIAN: All that tells me is that it is almost certain she was catheterized at that time, probably repeatedly with those procedures. You couldn't confirm it; it wouldn't help me because I would be willing to bet that she had been catheterized several times. So that the likelihood that she was infected 17 years ago between her stone, the operation for the stone and these others is high. There's smoke in the background but I've got 17 years of symptomless problems and, while the probability has gone up that this is all urinary infection, I'm fearful of being led down the primrose path and missing some more serious problem. And the price of missing a more serious abdominal disorder like some sort of perforating or penetrating ulcer or diverticulitis or something else or regional ileitis with some bladder symptoms is too high, so that I'm staying on guard.

Figure 9. The physician asks questions for several reasons. One is to assess the need for immediate action. This excerpt shows the clinician evaluating the danger of an acute condition requiring immediate intervention.

Discrepant Information. One of the problems that complicates the synthesis of clinical information is the need to deal with discrepant information. In some instances, new data conflict with data or hypotheses tentatively accepted. In the most obvious case, one fact simply contradicts another; but although it is generally easy to recognize such a situation, it is sometimes extremely difficult to resolve it. The clinicians invoked a number of indirect problem-solving strategies in order to determine which contradictory findings they would accept. The extent to which they were willing to investigate the matter depended largely on the consequences of accepting one or the other of these facts. If, despite their contradictory nature, the facts had relatively little impact on the current set of diagnostic hypotheses, the physician simply held in abeyance any further investigation of the contradiction. On the other hand, if accepting one or the other of these facts meant a gross change in the hypotheses concerning the patient's condition, then a more thorough investigation was undertaken.

In some instances, a new fact was discrepant only in the sense that it

---

DEALING WITH DISCREPANT INFORMATION

CLINICIAN: So now I've got to kind of revise my thinking a little bit and I've got to take into consideration that the albuminuria doesn't fit with the total picture nor does it fit with old chronic pyelonephritis, at least a 4 + if it's so. But if it's so, then I've got a problem. Pyelonephritis never, for practical purposes, produces that much proteinuria. If you told me she had a 1 or 2 + proteinuria even without much hematuria, I would say well she's had chronic pyelo and she's got recurrent pyelo now. It's been asymptomatic and she's entitled to 1 + albumin. At 2 + I become a little suspicious, 4 + it's out of the question. So I'm in the position of now having to believe that there's a mistake or having to believe that there are two diseases. She had every reason to have a urinary infection in terms of previous structural damage, of that there seems little doubt. But now it could turn out that she has some other underlying renal diseases that's evolved over the last 17 years unrelated to the predisposition resulting from the new renal disease and the structural abnormality which whatever it is will tend to increase the incidence of urinary infections. So, leaving out the current business of the pain and fever, I would be carrying two possible diagnoses in mind. One is pretty clearly infection and the other, What am I going to do about that proteinuria either in terms of finding out whether the test is wrong or else having to deal with it as a real finding.

Figure 10. Often the clinician must deal with ostensibly contradictory information. In some cases, a new fact requires the clinician to drop a current hypothesis or to reject the obvious interpretation of a fact. Here is an example of a physician trying to resolve such a conflict.

was inconsistent with a hypothesis that the clinician tentatively accepted, for example, when the hypothesis in question had been confirmed by an indirect method but no conclusive evidence for it had been obtained. In such a case, the acceptance of the new fact by the clinician required that he reject this hypothesis and develop an alternative one that included not only the previously gathered facts but the new fact as well. In some circumstances this was not difficult; but in others the cost of a revision was very great, and the clinician was more inclined to seek ways to explain why the discrepant fact could be rejected, why it could be considered an aberration, or even why it was consistent with his hypothesis after all. Figure 10 shows an example: The clinician felt strongly that chronic pyelonephritis was the correct diagnosis, but then he learned that a random sample of the patient's urine at an outside hospital had shown 4+ protein. Because heavy proteinuria is uncommon in patients with chronic pyelonephritis but common in other diseases—such as glomerulonephritis—the clinician considered the possibility of the presence of two independent renal diseases instead.

Discussion

In an era dominated by sophisticated diagnostic equipment and precise laboratory tests, the clinical axiom that diagnoses are most often made during the history-taking process has endured. Despite the respect afforded the cognitive skills that form the basis for this process, most of the standard sources used by students and house officers for guidance in taking the history of the present illness emphasize interviewing techniques and information acquisition rather than the principles and strategies of the diagnostic process. Chief concern in these texts is centered on the personal interaction with the patient, the need to avoid biased questions, the necessity of assessing the patient's reliability in recalling the history, and the value of thoroughly characterizing the patient's symptoms [1-10]. These issues have undeniable importance, but they are usually emphasized to the exclusion of the cognitive aspects of the diagnostic process. Critical elements such as how diagnostic possibilities are first introduced and evaluated, how competing diagnostic possibilities are eliminated, and what strategies should be used to obtain data with the greatest diagnostic information content are typically ignored.

The introspective reflection of experienced physicians on their own problem-solving methods has been the principal source of the literature on the diagnostic process. This literature has had considerable value in providing a useful framework for students and house officers learning how to approach a diagnostic problem. Additional evidence of the usefulness of introspection is derived from the recent demonstration that a computer program can be written that is based on the introspection of expert clincians and that closely simulates their diagnostic behavior [16]. Nonetheless, the analysis of protocols in this study has provided some insights that have not been described previously by investigators relying on introspective methods. Our observation that experienced physicians formulate diagnostic hypotheses early in the encounter with the patient has been made before [21], but the multiple clues that trigger these hypotheses and the importance of these initial hypotheses in providing a context in which the problem-solving process proceeds are not widely appreciated. Likewise, our observations on the process of hypothesis evaluation differ substantially from models of the diagnostic process described by others. One proposed model consists of an "algorithmic phase" in which a subset of the possible diagnosis is obtained after basic patient data are available [17]. Our observations suggest instead that both general and specific hypotheses are generated at a time when little information is available and that the hypotheses are progressively refined until they become more and more specific. Another investigator, relying on introspective techniques to refine the "problem-oriented" method, proposes that considerable clinical information should be accumulated without forming hypotheses; that "students must be taught to acquire a capacity for the 'sustained muddleheadedness' and a tolerance for ambiguity . . . so essential when difficult unexplained findings are dealt with. A diagnosis is a step forward only when it can be sustained by the evidence at hand" [22]. Our findings suggest that a clear-minded diagnostic context is essential to the tasks of hypothesis evaluation and information acquisition. Without this context, it is difficult to follow carefully and exhaustively those leads that make it possible to progressively refine preliminary diagnoses.

Research in Human Problem Solving

Our work follows the general line of investigation that can be termed the "study of human problem solving." Before 1950, investigations in psychology focused principally on perception and simple learning, especially in lower animals, but with an increasing emphasis on the study of these skills in humans. With a few exceptions, these studies of humans produced only the most rudimentary theories to "explain" the various intellectual abilities observed. More recently some psychologists began to devote greater attention to theories of human problem solving and have hypothesized information-processing mechanisms to explain some of the observed characteristics of human problem solving [23,24]. Developments in the study of language and visual perception contributed notably to this field as linquistic ability and perception were recognized to be rather complex problem-solving tasks [25].

Advances in computer technology provided perhaps the most powerful impetus to the development of information-processing theories of human problem solving, because computer programs constitute a medium in which such theories can be expressed precisely and in which the dynamic aspects of these theories can be tested. The earliest representation of theories of human problem solving as computer programs was accomplished for problems in logic [11] and the game of chess [12]. Research in the branch of computer science known as artificial intelligence contributed importantly because, for the first time, computer technology could be used to analyze problems that were neither highly structured nor narrowly circumscribed. Although the practical application of artificial intelligence techniques has been limited, sufficient evidence is already at hand that this discipline will enhance the development of problem-solving theories and practices in medicine [16].

The experimental design of modern attempts to develop information-processing theories of human problem-solving skills differs somewhat from that conventionally used. Because so many of the paths explored by an individual in an intellectual task are influenced both by his immediate experience and by his past experience with similar tasks, an analysis must be restricted to the problem-solving behavior of only a few individuals. However, this analysis is extremely detailed insofar as identifying what information the subject is using, how the information is acquired, and how the information is interpreted. Because of the richness of verbal data with respect to these questions, protocol analysis is often used in such investigations [13]. The observations from such analyses can then be subjected to study using the computer as a laboratory.

The most extensive use of protocol analysis has been in the study of the way people do certain highly structured symbolic tasks including proving theorems in logic, playing chess, solving cryptarithmetic problems, domains in which the dimensions of the task are sharply delineated [11–13]. Investigation of problem solving in these domains has shown several characteristics of the human problem solver—a limited capacity of short-term memory, a use of heuristic strategies to examine promising avenues, a tendency to search for information sequentially, and the importance of the problem solver's conceptualization of the problem at hand [13]. As noted earlier, all of these characteristics were also readily identified in our study.

As attention has turned recently to less highly structured problems, many researchers have confirmed the common-sense assertion that a large amount of knowledge, highly organized for the task at hand, is an essential ingredient in expert performance [26]. Without such knowledge, a person is forced by the limits of his cognitive abilities to pursue a rather plodding and often inefficient search for a solution to the problem before him. Our studies show the correctness of this view. All physicians used a large body of knowledge, both medical and nonmedical, in taking the history. Because of the nature of the case presented, the nephrologists were able to apply additional special knowledge in analyzing the problem, and therefore their problem-solving strategies appear to be more efficient. The other physicians, lacking this special knowledge, resorted more to methodical search.

Studies of problem solving by physicians have been few in number and limited in scope; nevertheless, several of our observations are closely analogous to the findings in these earlier studies. Using a variety of techniques, previous workers have identified the major mechanisms of hypothesis activation, hypothesis evaluation, and information acquisition [21,27–29]. Moreover, some of these studies have described the tendency of clinicians to immediately assess the gravity of the patient's illness and to evaluate the potential therapeutic implications [21,30]. Thus, our findings from the analysis of the diagnostic behavior of experienced clinicians appear to encompass a large fraction of these important principles and strategies.

Some Reflections on Methods

Although our study has provided new insights into the nature of clinical problem solving, it is appropriate to be cautious in interpreting any study based on the analysis of human behavior [31]. Statements that individuals make about their problem-solving behavior may be incomplete, misleading, or even incorrect. This problem is aptly described by some of the earlier researchers in this field: "On one level nothing seems complicated here—nothing is very different from a white rat sniffing his way through a maze. But still the problem persists that we are seeing only the superficial parts of the process—that there is a vast iceberg underneath concealed from our view and from the consciousness of the subject" [31]. Note should be made of possible artifacts induced during the experiment that may have influenced the outcome. Using a simulated patient probably has an additional effect. Moreover, the clinicians may have viewed the situation more as a game than a true-life experience. For example, if the experiment was viewed by the physician subjects as a competitive venture, they may have been influenced to "gamble"—to guess the right answer as soon as possible. Alternatively, these subjects might have suspected a trap (as they often do in discussing a case in a clinicopathologic conference) and therefore be more cautious. Finally, a clinician, by the process of articulating his problem-solving behavior, may have been influenced to alter his approach. As the physician discussed his reasons for asking a particular question or considering a particular hypothesis, secondary associations might have caused him to consider new approaches or new possibilities that he might otherwise not have considered.

Future Directions

It is important to acknowledge that the experiment reported here was limited in scope, because it was designed around a single simulated patient and a relatively small group of academic internists from one medical center. Certainly many other experiments can be conceived that involve more simulated patients, actual patients, a spectrum of diseases with a variety of complications, and a spectrum of physicians and students as experimental subjects. Even though the techniques of behavioral analysis in general and of this study in particular may have substantial shortcomings, we believe that useful information has been obtained. Experience with computer programs derived from the analysis of human behavior support this view in that the "trace" of the program rather closely resembles the protocol of the human subject in solving a given problem [16,31]. From experiences of this sort, some researchers have taken the view that the mystery associated with human problem solving is more illusory than actual. It has been pointed out that if indeed there are parts of the "iceberg" beneath the verbalized problem-solving behavior, then the subconscious parts of the process are no different in kind than the parts that the subject is able to articulate.

It would be useful to identify overall strategies of diagnostic reasoning so that ultimately we could submit them to test for efficiency. Such information would be of incalculable benefit in teaching the diagnostic process and in developing computer programs to take clinical histories. The preliminary results show that although the clinicians we studied used the same basic strategies for evaluating hypotheses and for collecting information, they often used them in different ways. One style recognized was that of the clinician who directed all his problem-solving efforts toward uncovering the core of the situation. In contrast, other clinicians approached the problem in a more methodical fashion, by engaging in a systematic exploration of a variety of aspects of the patient's condition. A third style was that of the clinician who probed a number of different directions as if he were hoping to uncover some important fact almost by chance. This clinician often interrupted one train of thought to jump to another area of investigation that superficially appeared to be rewarding. Finally, another approach is embodied by the clinician who began his analysis by going back as far in time as possible to obtain historical information, by dealing with the patient's problem from the point of view of its chronologic development, and by considering recent problems only after obtaining a clear vision of past events. Of course, the description of these styles is probably an oversimplification; nonetheless, preliminary observations suggest that the overall problem-solving strategies of experts may vary considerably. Our understanding of these gross diagnostic strategies is, at the moment, quite fragmentary.

Acknowledgment. The authors acknowledge the contribution of Peter B. Miller in the early phase of this project. Grant support: in part by grant 17629 from the National Heart, Blood, and Lung Institute to Baylor College of Medicine for a National Heart and Blood Vessel Research and Demonstration Center.

References

[1] Morgan WL Jr, Engel GL: *The Clinical Approach to the Patient.* Philadelphia, W.B. Saunders Co., 1969

[2] MacLeod JG (ed): *Clinical Examination.* New York, Churchill Livingstone, 1976

[3] Feinstein AR: *Clinical Judgment.* New York, Robert E. Krieger Publishing Co., 1967

[4] Bouchier IAD, Morris JS (eds): *Clinical Skills.* London, W.B. Saunders Co., 1976

[5] Stevenson I: *The Diagnostic Interview,* 2nd ed. New York, Harper & Row, Publishers, 1971

[6] Feinstein AR: An analysis of diagnostic reasoning. I. The domains and disorders of clinical macrobiology. *Yale J Biol Med* 46:212–232, 1973

[7] Wulff HR: *Rational Diagnosis and Treatment.* Oxford, Blackwell Scientific Publications, 1976

[8] DeGowin EL, DeGowin RL: *Bedside Diagnostic Examination,* 3rd ed. New York, Macmillan Publishing Co., Inc., 1976

[9] Enelow AJ, Swisher SN: *Interviewing and Patient Care.* New York, Oxford University Press, 1972

[10] Judge RD, Zuidema GD (eds): *Methods of Clinical Examination: A Physiologic Approach,* 3rd ed. Boston, Little, Brown and Co., 1974

[11] Newell A, Simon HA: The logic theory machine: a complex information processing system. *IRE Trans on Information Theory,* IT 2(3):61–79, 1956

[12] Newell A, Shaw JC, Simon HA: Chess-playing programs and the problem of complexity. *IBM J Res Dev* 2:320–335, 1958

[13] Newell A, Simon HA: *Human Problem Solving.* Englewood Cliffs, NJ, Prentice Hall, Inc., 1972

[14] Schwartz WB, Gorry GA, Kassirer JP, Essig A: Decision analysis and clinical judgment. *Am J Med* 55:459–472, 1973

[15] Pierce JR: *Symbols, Signals and Noise: The Nature and Process of Communication.* New York, Harper Torchbooks, 1961

[16] Pauker SG, Gorry GA, Kassirer JP, Schwartz WB: Towards the simulation of clinical cognition. Taking a present illness by computer. *Am J Med* 60:981–996, 1976

[17] DeDombal FT: Computer-assisted diagnosis, in *Principles and Practice of Medical Computing,* edited by Whitby LG, Lutz W, Edinburgh, Livingstone, 1971

[18] Shortliffe EH, Axline SG, Buchanan BG, Merigan TC, Cohen SN: An artificial intelligence program to advise physicians regarding antimicrobial therapy. *Comput Biomed Res* 6:544–560, 1973

[19] Kulikowski CA, Weiss S, Safir A: Glaucoma diagnosis and therapy by com-

puter, in *Proceedings of the Annual Meeting of the Association for Research in Vision and Ophthalmology,* Sarasota, Florida, May 1973

[20] Pople H, Werner G: An information processing approach to theory formation in biomedical research, in *Proc AFIPS Spring Joint Computer Conf,* 1972, p. 1125

[21] Elstein AS, Kagan N, Shulman LS, Jason H, Loupe MJ: Methods and theory in the study of medical inquiry. *J Med Educ* 47:85–92, 1972

[22] Weed LL: Medical records that guide and teach. *N Engl J Med* 278:593–600, 1968

[23] Miller GA: The magical number seven, plus or minus two: some limits on our capacity for processing information. *Psychol Rev* 63:81–97, 1956

[24] Bruner JS, Goodnow JJ, Austin GA: *A Study of Thinking.* New York, Wiley, John, & Sons, Inc., 1956

[25] Chomsky AN: Three models for the description of language. *See* reference 11, pp. 113–124

[26] Winston PH: *Artificial Intelligence.* Reading, Addison-Wesley, 1977

[27] Buckingham WB, Sparberg M, Brandfonbrener M: *A Primer of Clinical Diagnosis.* New York, Harper & Row, Publishers, 1971

[28] Dudley HAF: Clinical method. *Lancet* 1:35–37, 1971

[29] Dudley HAF: The clinical task. *Lancet* 2:1352–1354, 1970

[30] DeDombal FT, Horrocks JC, Staniland JR, Guillou PJ: Production of artificial "case histories" by using a small computer. *Br Med J* 2:578–581, 1971

[31] Simon HA: *The New Science of Management Decision.* New York, Harper & Row, Publishers, 1960

Towards the Simulation of Clinical Cognition: Taking a Present Illness by Computer

Stephen G. Pauker, G. Anthony Gorry, Jerome P. Kassirer, and William B. Schwartz

Remarkably little is known about the cognitive processes which are employed in the solution of clinical problems. This paucity of information ·is probably accounted for in large part by the lack of suitable analytic tools for the study of the physician's thought processes. Here we report on the use of the computer as a laboratory for the study of clinical cognition.

Our experimental approach has consisted of several elements. First, cognitive insights gained from the study of clinicians' behavior were used to develop a computer program designed to take the present illness of a patient with edema. The program was then tested with a series of prototypical cases, and the present illnesses generated by the computer were compared to those taken by the clinicians in our group. Discrepant behavior on the part of the program was taken as a stimulus for further refinement of the evolving cognitive theory of the present illness. Corresponding refinements were made in the program, and the process of testing and revision was continued until the program's behavior closely resembled that of the clinicians.

The advances in computer science that made this effort possible include "goal-directed" programming, pattern-matching and a large associative memory, all of which are products of research in the field known as "artificial intelligence." The information used by the program is organized in a highly connected set of associations which is used to guide such activities as checking the validity of facts, generating and testing hypotheses, and constructing a coherent picture of the patient. As the program pursues its interrelated goals of information gathering and diagnosis, it uses knowledge of diseases and pathophysiology, as well as "common sense," to dynamically assemble many small problem-solving strategies into an integrated history-taking process.

We suggest that the present experimental approach will facilitate accomplishment of the long-term goal of disseminating clinical expertise *via* the computer.

Reprinted with permission from *The American Journal of Medicine, 60,* pp. 981—996. Copyright 1976 by The Yorke Medical Group.

During the last decade there has been increasing interest in the use of the computer as an aid to both clinical diagnosis and management. Programs have been written that can carry out a review of systems [1], guide in the evaluation of acid-base disorders [2,3], recommend the appropriate dose of digitalis [4,5], and weigh the risks and benefits of alternative modes of treatment [6]. Some of these programs have been used to a limited extent in clinical practice, whereas others are prototypes which, although not yet of practical value, offer promise for the future. All, however, have the underlying characteristics that they are highly structured and that they deal with well-defined, sharply-constrained problems. In nearly all instances, the use of a formalism, such as a flow chart [1–3], decision analysis [6], or a mathematical algorithm [4,5] is the guiding principle used to capture clinical expertise in the computer.

There are, however, aspects of clinical medicine that cannot be reduced to formalisms, i.e., situations in which a fixed recipe cannot provide the skilled guidance of the experienced clinician. To deal with this class of problems, new and more flexible strategies are under development, but work on such strategies is still in its embryonic phase [7–9].

In this paper, we report on the development of a computer program that uses unstructured problem-solving technics to take the history of the present illness of a patient with edema.[1] We have chosen the "present illness" for investigation, because it is prototypical of clinical problems that demand complex problem-solving strategies. The present illness is, furthermore, the keystone upon which a physician builds his diagnosis and upon which he bases many of his subsequent management decisions. Although we have examined only a limited range of issues in the present illness, we believe that our effort is a first step towards a full understanding of the way in which a physician carries out the history-taking process.

Computer Science in the Study of Clinical Cognition

Our attempt to simulate the unstructured problem-solving processes of the present illness falls into the domain of computer science known as **artificial intelligence.** Research in this field is concerned with producing computer programs which exhibit behavior that would be termed "intelligent," if such behavior were that of a person. Examples of such work are programs that, to a limited extent, understand English, make sense of certain kinds of visual scenes and control the operations of robots [11]. Such research has been underway for 20 years [12] and, during this time, some major lessons have been learned. Perhaps the most important dis-

---

[1]The program described here should be contrasted with the well bounded "present illness algorithms" [10], which rely on flow charts for their implementation and which refer the patient to the physician for further questioning whenever the situation appears to be complex or serious.

covery has been that formalisms alone, e.g., cybernetics [13], mathematical logic [14] and information theory [15] cannot produce intelligent behavior in complex, "real-world" situations. It has become abundantly clear that no single, formal approach can accommodate the knowledge of first principles and the experience, common sense[2] and guesswork [16] required for "intelligent" activities.

Because of the obvious competence of people to carry out activities that formalisms cannot, artificial intelligence researchers have turned more recently to the study of human problem-solving [11,17]. The study of natural intelligence, in fact, has become the central activity of artificial intelligence, and the experimental method of the field now emphasizes the use of computer systems as laboratories in which theories of human problem-solving can be represented and tested [18].

Conventional computer programming concepts and structures have proved inadequate to express complex theories of human problem-solving; however, new technics have been developed that ameliorate these technologic difficulties. Greatly improved systems have been created for managing very large collections of facts, and new "goal-directed" programming languages have been designed for utilizing these facts in the solution of difficult problems. Through the appropriate statement of "goals," it is possible to construct a program that brings knowledge to bear *when it is required.* As new facts are obtained, such programs can dynamically organize many small problem-solving technics into a coherent strategy which can respond flexibly to the changing picture of the world.[3] Equally important, as we shall discuss, these new languages provide means for giving a program *advice* as to when a particular piece of knowledge may be useful and how that knowledge should be applied to particular situations.

We believe that the ideas and technology now emerging from artificial intelligence research should make possible realistic simulations of human

---

[2]By the term *common sense,* we mean all the ordinary, rather pedestrian knowledge about everyday occurrences possessed by reasonably intelligent people.

[3]Expressed in technical terms, these languages do not require a detailed, rigid program because of *pattern-directed invocation.* Each subroutine contains a statement of what it potentially can accomplish, so the programmer need not specify which subroutines (or even that any subroutine) should carry out a desired action. Rather, he can specify the desired effect or goal and ask the computer to identify and use those subroutines that appear relevant. This type of program organization has many applications, such as offering heuristic advice and generating hypotheses. As an example of one class of problems which is very difficult to solve with conventional technics, but which is trivial with this type of language, consider the problem of logical deduction. The program is told "All Greeks are poets" and "Anyone born in Athens is a Greek." We then tell the program that "Constantine was born in Athens" and ask "is Constantine a poet?". The program *automatically* deduces the answer, basically using the same process which we would. That is, it sets about to find out if Constantine is a poet. It realizes that one way to answer this question is to determine if he is a Greek, and therefore it asks if Constantine was born in Athens. When it discovers that he was, the original question is answered.

problem-solving strategies. In assessing the feasibility of building any "intelligent" program, however, some vital questions must be answered: "What is expert knowledge?," "How much knowledge is required?," "How should it be organized and how should it be applied?." The answers to these questions will come only from the careful study of real problem domains, and the success of such studies will be determined in large part by the *boundedness* of the problem domain under consideration. We believe that medicine, with its highly-developed taxonomy, its codified knowledge-base, the generally repetitive nature of the problem-solving encounters and the existence of acknowledged experts, constitutes a promising problem domain because of its relatively well-bounded character. We therefore believe that, building on the technology at hand, acceptable progress can be made toward the development of sophisticated programs that can deal competently with complex clinical problems. To achieve such progress, however, the essential first step is to examine in depth the nature of the clinician's cognitive processes.

Methods of Procedure

Our first efforts were directed towards elucidating a number of the problem-solving strategies that physicians use in taking the history of the present illness of a patient with edema. This analytic effort was carried out through introspection and through direct observations of clinicians' problem-solving behavior. The insights gained in this way were represented as a computer program (using the CONNIVER programming system [19]) which incorporates the goal-directed technics described in the previous section. The program was then tested with a series of prototypical cases in which edema was the presenting problem, and the questioning strategy followed by the program was compared to that of the physicians whom it was intended to simulate.

It immediately became apparent that the program's behavior differed markedly from that of the physicians, but, by examining specific discrepancies, we were able to recognize components of the clinicians' reasoning process that had been misunderstood or neglected in our initial analysis. With these new insights, we revised the program and evaluated its history-taking performance again. With each iteration of this process, the performance of the program improved and our insights into the cognitive process deepened. The study was terminated when the program closely simulated the manner in which the physician members of the team take the present illness of a patient with edema.[4]

---

[4]Our program simulates the behavior of the particular physicians in our group. The question of differences in problem-solving behavior among physicians is one which we intend to pursue further.

Examples of Computer-Generated Present Illnesses

Figure 1 presents a portion of a typical dialogue between user (a physician) and program. The language of both the questions and the computer-generated summaries (Figures 2 through 5) is rather stilted because most of our effort has been devoted to examining the history-taking process, not to producing a polished output. Each case demonstrates the program's use of somewhat different over-all strategies of history-taking.

PRESENTING PROBLEM: A YOUNG MAN WITH MASSIVE PEDAL EDEMA

1A   IS PEDAL EDEMA, WHICH IS MASSIVE,
       1.   FOR-DAYS?
       2.   FOR-WEEKS?
       3.   FOR-MONTHS?
       4.   FOR-YEARS?   >>1

1B   IS IT,
       1.   FIRST-TIME?
       2.   INFREQUENT?
       3.   OCCASIONAL?
       4.   FREQUENT?   >>1

1C   IS IT,
       1.   WITHOUT-TEMPORAL-PATTERN?
       2.   WORSE-IN-MORNING?
       3.   WORSE-IN-EVENING?   >>3

1D   IS IT,
       1.   SYMMETRICAL?
       2.   ASYMMETRICAL?   >>1

1E   IS IT,
       1.   PITTING?
       2.   NON-PITTING?   >>1

1F   IS IT,
       1.   ERYTHEMATOUS?
       2.   NOT ERYTHEMATOUS?   >>2

1G   IS IT,
       1.   PAINFUL?
       2.   NOT PAINFUL?   >>2

2   WELL, IS THERE DYSPNIA?   >>NO

Figure 1. Case 1. Initial portion of a representative present illness dialogue. The questions were asked by the computer and the responses (preceded by >>) are those of the user. The complete present illness, as generated by the computer, is shown in Figure 2.

3	BY THE WAY, IS THERE HEAVY ALCOHOL CONSUMPTION?	≫NO

3A   IS ALCOHOL CONSUMPTION,
1. NONE?
2. RARE?
3. SOCIAL?   ≫3

4   TELL ME, IS THERE JAUNDICE?   ≫NO

5   TELL ME, WAS THERE PROTEINURIA IN THE PAST?   ≫UNKNOWN

5A   IS THERE LARGE-POLICY LIFE INSURANCE?   ≫NO

5A1   IS LIFE INSURANCE,
1. NONE?
2. UNABLE-TO-OBTAIN?
3. SMALL-POLICY?   ≫3

5B   IS THERE SERVED-IN ARMED FORCES?   ≫YES

6   WELL, IS THERE VARICOSE VEINS?   ≫NO

7   WAS THERE EDEMA IN THE PAST?   ≫NO

8   WELL, IS THERE ANY HEMATURIA?   ≫NO

8A   IS THERE . . . .

Figure 1. Continued.

Case 1

Figure 2 shows the computer-generated summary of Case 1, a patient with **idiopathic nephrotic syndrome.** The computer was given as the chief complaint: "a young man with massive pedal edema." The behavior of the program can be briefly summarized as follows. The computer characterized the edema in detail and, in light of the specific findings, turned to questions designed to elucidate etiology. After quickly determining that there was no history suggestive of congestive heart failure, alcoholic cirrhosis, varicosities or renal failure, it noted that the patient had several findings strongly suggestive of nephrotic syndrome. The program then initiated a search for causes of the nephrotic syndrome, first exploring the possibility that the patient was suffering from poststreptococcal glomerulonephritis and then looking for evidence of a systemic disease such as lupus erythematosus. Finding no evidence of a systemic disorder, the program made the diagnosis of nephrotic syndrome, probably idiopathic in character, but indicated that acute glomerulonephritis remained as a second, albeit much less likely, possibility. Note incidentally, that the

PRESENTING PROBLEM: A YOUNG MAN WITH MASSIVE PEDAL EDEMA.

THE CASE CAN BE SUMMARIZED AS FOLLOWS:

THIS IS A YOUNG MAN WHO HAS PEDAL EDEMA WHICH IS NOT PAINFUL, NOT ERYTHEMATOUS, PITTING, SYMMETRICAL, WORSE-IN-EVENING, FIRST-TIME, FOR-DAYS AND MASSIVE. HE DOES NOT HAVE DYSPNEA. HE HAS SOCIAL ALCOHOL CON-SUMPTION. HE DOES NOT HAVE JAUNDICE. IT IS NOT EXPLICITLY KNOWN WHETHER IN THE PAST HE HAD PROTEINURIA, BUT HE HAS SMALL-POLICY LIFE INSURANCE, AND HE HAS SERVED IN ARMED FORCES. HE DOES NOT HAVE VARICOSE VEINS. IN THE PAST HE DID NOT HAVE EDEMA. HE DOES NOT HAVE HEMATURIA. HE HAS NORMAL BUN. HE HAS NORMAL CREATININE. HE HAS PERI-ORBITAL EDEMA, WHICH IS WORSE-IN-MORNING, FIRST-TIME, FOR-DAYS AND SYMMETRICAL. HE HAS LOW SERUM ALBUMIN CONCENTRATION. HE HAS HEAVY PROTEINURIA, WHICH IS > 5GRAMS/24HRS. HE HAS MODERATELY-ELEVATED, RISING WEIGHT. IN THE RECENT PAST HE DID NOT HAVE PHARYNGITIS. IN THE RECENT PAST HE HAD NOT ATTENDED SCHOOL. IN THE RECENT PAST HE HAD NOT ATTENDED SUMMER CAMP. IN THE RECENT PAST HE HAD NOT BEEN EXPOSED TO STREPTOCOCCI. IN THE RECENT PAST HE DID NOT HAVE FEVER. IT IS SAID, BUT HAS BEEN DISREGARDED, THAT HE HAS RED-CELL-CASTS-IN URINARY SEDIMENT. HE DOES NOT HAVE JOINT PAIN. HE DOES NOT HAVE RASH. HE HAS NEGATIVE ANA. HE DOES NOT HAVE FEVER. HE HAS NOT RECEIVED ANTIBIOTIC. HE DOES NOT HAVE ANEMIA. IN THE PAST HE DID NOT HAVE HEMATURIA.

DIAGNOSES THAT HAVE BEEN *ACCEPTED* ARE: NEPHROTIC SYNDROME AND SODIUM RETENTION.

THE LEADING HYPOTHESIS IS IDIOPATHIC NEPHROTIC SYNDROME.

HYPOTHESES BEING CONSIDERED:

	A. fit of case to hypothesis	B. fraction of findings explained by hypothesis	average of A and B
IDIOPATHIC NEPHROTIC SYNDROME	0.80	0.37	0.58
ACUTE GLOMERULONEPHRITIS	0.22	0.27	0.24
HENOCH-SCHOENLEIN PURPURA	0.07	0.10	0.09

Figure 2. Case 1. Computer-generated present illness of a patient with idiopathic nephrotic syndrome. Note that the diagnosis was not available to the computer; the program was provided only with the description of the presenting problem. The rank-listing at the bottom of the figure is based on the average of the "fit" of the case to the hypothesis (column A) and the fraction of findings explained by the hypothesis (column B). For details of the evaluation (scoring) procedure, see text.

program disregarded the statement that red cell casts had been seen because it concluded that in the absence of hematuria the report of red cell casts was almost certainly in error. Also note that the questions about life insurance and military service were utilized to determine if protein-uria had been present sometime in the past.

Case 2

Figure 3 summarizes the present illness of a patient with **Laennec's cirrhosis**. The computer was given as a chief complaint: "a woman with pedal edema." In response, it obtained a detailed description of the character of the edema and then undertook an exploration of possible etiol-

---

PRESENTING PROBLEM: A MIDDLE AGED WOMAN WITH PEDAL EDEMA.

THE CASE CAN BE SUMMARIZED AS FOLLOWS:

THIS IS A MIDDLE-AGED WOMAN, WHO HAS PEDAL EDEMA, WHICH IS NOT-PAIN-FUL, NOT-ERYTHEMATOUS, PITTING, SYMMETRICAL, 4+, WITHOUT-TEMPORAL-PATTERN, OCCASIONAL AND FOR-WEEKS. SHE DOES NOT HAVE DYSPNEA. SHE HAS HEAVY ALCOHOL CONSUMPTION. SHE HAS JAUNDICE. SHE HAS PAINFUL HEPATO-MEGALY. SHE HAS SPLENOMEGALY. SHE HAS ASCITES. SHE HAS PALMAR ERY-THEMA. SHE HAS SPIDER ANGIOMATA. SHE DOES NOT HAVE PAROTID ENLARGE-MENT. SHE HAS MODERATELY-ELEVATED, DIRECT-AND-INDIRECT BILIRUBIN. SHE HAS PROLONGED PROTHROMBIN TIME. SHE HAS MODERATELY-ELEVATED SGPT. SHE HAS MODERATELY-ELEVATED SGOT. SHE HAS MODERATELY-ELEVATED LDH. SHE HAS NOT-RECEIVED BLOOD TRANSFUSIONS. SHE HAS NOT-EATEN CLAMS. SHE DOES NOT HAVE ANOREXIA. SHE HAS MELENA. SHE DOES NOT HAVE HEMATEME-SIS. SHE HAS LOW SERUM IRON. SHE HAS ESOPHAGEAL VARICES.

DIAGNOSES THAT HAVE BEEN *ACCEPTED* ARE: ALCOHOLISM AND GI BLEEDING.

THE LEADING HYPOTHESIS IS CIRRHOSIS.

HYPOTHESES BEING CONSIDERED:

	A. fit of case to hypothesis	B. fraction of findings explained by hypothesis	average of A and B
CIRRHOSIS	0.72	0.78	0.75
HEPATITIS	0.75	0.30	0.53
PORTAL HYPERTENSION	0.72	0.17	0.45
CONSTRICTIVE PERICARDITIS	0.17	0.13	0.15

Figure 3. Case 2. Computed-generated present illness of a patient with cirrhosis of the liver. The format is identical to that of Figure 2.

ogies. Upon finding that the patient drank large quantities of alcohol, it turned to cirrhosis as a working hypothesis and quickly uncovered many stigmata of liver disease. The program also briefly explored other etiologies of liver disease, such as the hepatitis induced by transfusions or by the ingestion of raw shellfish, but could find no evidence in support of these diagnoses. It then returned to the primary hypothesis of cirrhosis and, in searching for possible complications, noted the presence of both esophageal varices and chronic gastrointestinal bleeding. It concluded that the patient had alcoholic cirrhosis and that hepatitis was an alternative, but much less likely, possibility.

Case 3

Figure 4 shows the computer-generated summary of a patient with **acute tubular necrosis** produced by carbon tetrachloride exposure. The computer was given as a chief complaint: "a young man with edema and oliguria." The program immediately undertook a search for causes of acute renal failure. It first focused on the diagnosis of acute glomerulonephritis but could find no evidence of streptococcal exposure. It next explored the possibility of acute tubular necrosis, but was unable to find an etiologic factor. When the program later assessed the characteristics of the urine sediment, it noted, however, many hallmarks of tubular injury. Pursuing this lead, it soon uncovered an exposure to a cleaning fluid which it presumed contained carbon tetrachloride. It then explored the possibility that acute hypotension had also contributed to the development of the oliguria but could obtain no evidence in support of this hypothesis. Finally it determined that body weight was increasing and from this fact concluded that the patient was retaining sodium. Because the data base does not currently include the distinction between the retention of salt and the retention of free water, the program could not arrive at the correct interpretation of the weight gain, namely, that the overhydration was due to water retention per se.

Case 4

Figure 5 gives the present illness of a man with **constrictive pericarditis** secondary to tuberculosis. The program was given as the chief complaint: "a middle-aged man with ascites and pedal edema." After further characterizing the edema, the computer focused on an hepatic etiology and found that the patient had an enlarged liver. Although subsequent questioning revealed only "social" alcohol consumption, the program persisted in its search for stigmata of cirrhosis. When none was found, it turned to a possible cardiac etiology and noted that chest pain was a prominent complaint; the pain was, however, more characteristic of pleu-

PRESENTING PROBLEM: A YOUNG MAN WITH PEDAL EDEMA AND OLIGURA.

THE CASE CAN BE SUMMARIZED AS FOLLOWS:

THIS IS A YOUNG MAN, WHO HAS OLIGURIA. HE HAS PEDAL EDEMA, WHICH IS NOT-PAINFUL, NOT-ERYTHEMATOUS, PITTING, SYMMETRICAL, 3+, WITHOUT-TEMPORAL-PATTERN, FIRST-TIME AND FOR-DAYS. IT HAS BEEN DENIED THAT HE HAS RECENT SCARLET FEVER. IN THE RECENT PAST HE DID NOT HAVE PHARYNGITIS. IN THE RECENT PAST HE HAD NOT-ATTENDED SCHOOL. IN THE RECENT PAST HE HAD NOT-ATTENDED SUMMER CAMP. IN THE RECENT PAST HE HAD NOT-BEEN-EXPOSED-TO STREPTOCOCCI. HE HAS NOT-RECEIVED RADIOGRAPHIC CONTRAST MATERIAL. HE HAS NOT-RECEIVED NEPHROTOXIC DRUGS. IN THE RECENT PAST HE DID NOT HAVE HYPOTENSION. HE HAS MODERATELY-ELEVATED URINE SODIUM. HE HAS URINE SPECIFIC GRAVITY WHICH IS ISOSTHENURIC. HE HAS NO-RED-CELLS-IN, NO-WHITE-CELLS-IN, RENAL-CELLS-IN, NO-RENAL-CELL-CASTS-IN, HYALINE-CASTS-IN URINARY SEDIMENT. IT IS NOT EXPLICITLY KNOWN WHETHER HE HAS BEEN-EXPOSED-TO CARBON TETRACHLORIDE, BUT HE HAS BEEN-EXPOSED-TO A CLEANING FLUID. HE DOES NOT HAVE HYPOTENSION. HE DOES NOT HAVE TACHYCARDIA. HE HAS NORMAL-TURGOR-AND-PERFUSION-OF SKIN. HE HAS MODERATELY-ELEVATED, RISING WEIGHT.

DIAGNOSES THAT HAVE BEEN *ACCEPTED* ARE: SODIUM RETENTION, EXPOSURE TO NEPHROTOXINS, EXPOSURE TO HEPATOTOXINS AND ACUTE RENAL FAILURE.

THE LEADING HYPOTHESIS IS ACUTE TUBULAR NECROSIS.

HYPOTHESES BEING CONSIDERED:

	A. fit of case to hypothesis	B. fraction of findings explained by hypothesis	average of A and B
ACUTE TUBULAR NECROSIS	0.50	0.37	0.43
ACUTE GLOMERULONEPHRITIS	0.20	0.21	0.20
IDIOPATHIC NEPHROTIC SYNDROME	0.18	0.16	0.17
CHRONIC GLOMERULONEPHRITIS	0.19	0.11	0.15

Figure 4. Case 3. Computer-generated present illness of a patient with acute tubular necrosis. The format is identical to that of Figure 2.

ral or pericardial than of myocardial disease. It next found that there was both neck vein distention and orthopnea but that there was no paroxysmal nocturnal dyspnea. These clinical findings, in combination with ascites, suggested the diagnosis of pericardial disease. Further questioning then revealed many of the stigmata of constrictive pericarditis.

Because even the experienced clinician often confuses constrictive pericarditis with cirrhosis, it is understandable why the diagnosis of cirrhosis

PRESENTING PROBLEM: A MIDDLE AGED MAN WITH ASCITES AND PEDAL EDEMA.

THE CASE CAN BE SUMMARIZED AS FOLLOWS:

THIS IS A MIDDLE-AGED MAN WHO HAS ASCITES. HE HAS PEDAL EDEMA, WHICH IS NOT-PAINFUL, NOT-ERYTHEMATOUS, PITTING, SYMMETRICAL, 4+, WORSE-IN-EVENING, OCCASIONAL AND FOR-MONTHS. HE HAS SOCIAL ALCOHOL CONSUMPTION. HE HAS HEPATOMEGALY. HE DOES NOT HAVE JAUNDICE. HE DOES NOT HAVE PALMAR ERYTHEMA. HE DOES NOT HAVE SPIDER ANGIOMATA. HE DOES NOT HAVE PAROTID ENLARGEMENT. HE DOES NOT HAVE GYNECOMASTIA. HE DOES NOT HAVE TESTICULAR ATROPHY. HE HAS NORMAL BILIRUBIN. HE HAS NORMAL PROTHROMBIN TIME. HE HAS NORMAL SGPT. HE HAS NORMAL SGOT. HE HAS CHEST PAIN WHICH IS RELIEVED-BY-SITTING-UP, WITHOUT-RADIATION, MODERATE, OCCASIONAL, FOR-SECONDS AND SHARP. HE HAS EXERTIONAL DYSPNEA. HE HAS ORTHOPNEA. HE DOES NOT HAVE PAROXYSMAL NOCTURNAL DYSPNEA. HE HAS ELEVATED NECK VEINS. HE HAS KUSSMAUL'S SIGN. HE HAS PERICARDIAL KNOCK. HE HAS DISTANT HEART SOUNDS. HE HAS PERICARDIAL-CALCIFICATION-ON, NORMAL-HEART-SIZE-ON, CLEAR-LUNG-FIELDS-ON CHEST XRAY.

THE LEADING HYPOTHESIS IS CONSTRICTIVE PERICARDITIS.

HYPOTHESES BEING CONSIDERED:

	A. fit of case to hypothesis	B. fraction of findings explained by hypothesis	average of A and B
CONSTRICTIVE PERICARDITIS	0.78	0.50	0.64
CONGESTIVE HEART FAILURE	0.44	0.21	0.32

Figure 5. Case 4. Computer-generated present illness of a patient with constrictive pericarditis. The format is identical to that of Figure 2.

was pursued with such vigor. Note, however, that the program was deficient in that it failed to explore other etiologies of predominantly right-sided cardiac failure, such as cor pulmonale and multiple pulmonary emboli; this short-coming is explained by the fact that the current data base does not include information about these latter diagnoses.

Nature of the Underlying Computer Programs

In this section we shall first discuss the over-all behavior of the program in terms of its major components and the way that these components interact. We shall then consider in detail the underlying processes used by the program.

An Overview of the Present Illness Program

In taking a history of the present illness, the program, much like the physician, tries to develop a sufficient "understanding" of the patient's complaints to form a reasonable basis on which to evaluate the clinical problem and to lay the ground-work for subsequent management decisions. It accomplishes this goal by undertaking two processes, information-gathering and diagnosis. Although these two threads of the problem-solving process are interwoven, for clarity of exposition we shall consider them separately.

By **information gathering,** we mean the accumulation of a profile of data concerning the patient. Because there are innumerable "facts" which could be gathered, one needs a sharp focus for this activity. This focus is obtained through the pursuit of a small set of diagnostic hypotheses that are suggested by the presenting complaints.

The process of **diagnosis,** in contrast, is an attempt to infer the meaning of a constellation of given findings and does not involve the acquisition of additional information about the patient; rather, it is concerned with the processing of the available facts. When additional findings are required, the diagnostic process turns again to the information gathering process. Thus, the history taking process is directed both at establishing *what the facts are* and at establishing *what the facts mean* [20].

In taking a present illness, our program uses the chief complaint to generate hypotheses about the patient's condition. It also actively seeks additional clinical information to accomplish a number of different tasks including testing hypotheses and eliminating unlikely ones. Any of these activities may spawn further tasks, such as checking the validity of a newly discovered fact or asking about related findings. As will become evident, however, this brief description understates both the complexity of the program's behavior and the differences between this program and others previously reported.

The Basic Components of the Program

The complexity of the program's behavior is the result of the interaction of the four factors schematically shown in Figure 6: (1) the *Patient Specific Data,* (2) the *Supervisory Program,* (3) the *Short-Term Memory* and (4) the *Associative Memory.*

The Patient Specific Data. These are the facts provided by the user either spontaneously or in response to questions asked by the program. These data comprise the computer's knowledge about the patient.

The Supervisory Program. The supervisory program guides the computer in taking the present illness and oversees the operation of various sub-processes, such as selecting questions, seeking and applying relevant

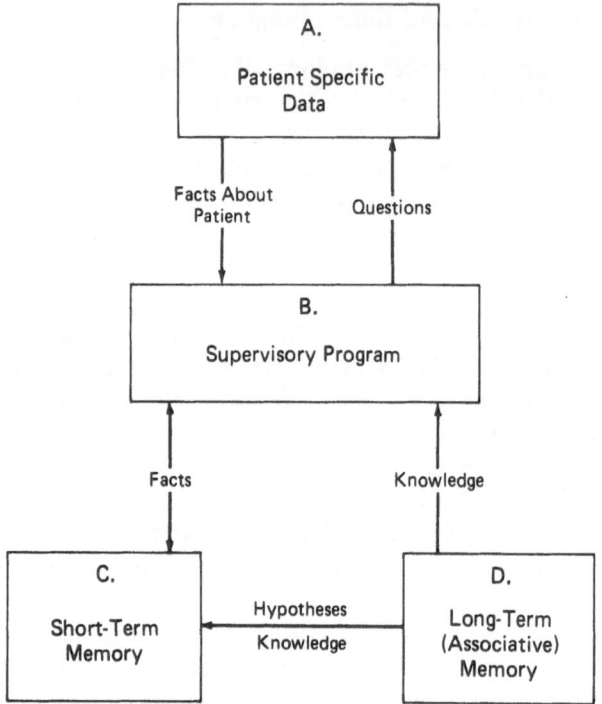

Figure 6. Overview of computer program organization. Clinical data (A) are presented to the supervisory program (B), which places them in short-term memory (C). The supervisory program, after consulting both short-term (C) and long-term memories (D), generates hypotheses and moves the information associated with these hypotheses from long-term to short-term memory. The supervisory program then asks for additional patient-specific data relevant to its hypotheses. At every stage, each hypothesis is evaluated (scored) by the program to determine whether it should be rejected, accepted or considered further.

advice, and processing algorithms (such as flow charts). The principal goal of this supervisor is to arrive at a coherent formulation of the case, by quickly generating and testing hypotheses and by excluding competing hypotheses. At the present time, there are about 300 potential questions which relate to over 150 different concepts that the program can employ in its information-gathering activities.

The Short-Term Memory. The short-term memory is the site in which data about the patient interact with general medical knowledge that is kept in long-term memory (vide infra). The supervisory program determines which aspects of this general knowledge enter the short-term memory and how such knowledge is melded with the patient-specific data that are under consideration. The amount of information in short-term memory is quite variable, depending on the complexity of the case and the number of active hypotheses. For a simple case, the short-term memory might contain only two or three hypotheses, and the knowledge and deductions associated with them. In a complex or puzzling case, it might contain five or ten hypotheses.

The Associative (Long-Term) Memory. The long-term memory contains a rich collection of knowledge, organized into packages of closely related facts called **frames** [16]. Frames are centered around diseases (such as acute glomerulonephritis), clinical states (such as nephrotic syndrome) or physiologic states (such as sodium retention). Within each frame is a rich knowledge structure which includes prototypical findings (signs, symptoms, laboratory data), the time course of the given illness, and rules for judging how closely a given patient might "match" the disease or state which the frame describes. A typical example of a frame (nephrotic syndrome) is shown in Figure 7.

As shown in Figure 8, the frames are linked into a complex **network**. In the figure each frame is represented as a colored sphere (diseases are dark gray, clinical states are light gray, and physiologic states are medium gray), and the links between the frames are represented as labeled rods. These links depict a variety of relations, such as "may be caused by" and "may be complicated by."

In addition to information about diseases and physiology, the network contains **knowledge of the real world.** This information is also organized into frames and is linked to areas of the associative memory in which such common-sense knowledge is relevant.

The present program contains over 70 frames related to some 20 different diseases and to a variety of clinical and physiologic states that are associated with these diseases. Frames typically contain five to ten findings, three or four exclusionary rules, ten to twenty scoring parameters and five to ten links to other frames in the network. Because the frames are presented to the computer as separate descriptions, which the program links into the network, the addition of frames to the system is a relatively simple task.

The Operation of the Program

In this section we shall consider in detail the individual processes by which the program combines patient-specific data and knowledge from the associative memory to produce the behavior shown in the illustrative cases. Basically, the program alternates between asking questions to gain new information and integrating this new information into a developing picture of the patient. A typical cycle consists of (1) characterizing findings, (2) seeking advice on how to proceed, (3) generating hypotheses, (4) testing hypotheses and (5) selecting questions.

NAME: NEPHROTIC SYNDROME

IS-A-TYPE-OF: CLINICAL STATE
FINDING: LOW SERUM ALBUMIN CONCENTRATION
FINDING: HEAVY PROTEINURIA
FINDING: >5GRAMS/24HRS PROTEINURIA
FINDING: MASSIVE, SYMMETRICAL EDEMA
FINDING: EITHER FACIAL OR PERI-ORBITAL.
                        AND SYMMETRICAL EDEMA
FINDING: HIGH SERUM CHOLESTEROL CONCENTRATION
FINDING: URINE LIPIDS PRESENT
MUST-NOT-HAVE:
    PROTEINURIA ABSENT
IS-SUFFICIENT:
    BOTH MASSIVE EDEMA AND >5GRAMS/24HRS PROTEINURIA
MAJOR-SCORING:
    SERUM ALBUMIN CONCENTRATION
        LOW: 1.0
        HIGH: −1.0
    PROTEINURIA:
        >5GRAMS/24HRS: 1.0
        HEAVY: 0.5
        EITHER ABSENT OR LIGHT: −1.0
    EDEMA:
        MASSIVE AND SYMMETRICAL: 1.0
        NOT MASSIVE BUT SYMMETRICAL: 0.3
        ERYTHEMATOUS: −0.2
        ASYMMETRICAL: −0.5
        ABSENT: −1.0
MINOR-SCORING:
    SERUM CHOLESTEROL CONCENTRATION:
        HIGH: 1.0
        NOT HIGH: −1.0
    URINE LIPIDS:
        PRESENT: 1.0
        ABSENT: −0.5
MAY-BE-CAUSED-BY:
    ACUTE GLOMERULONEPHRITIS,
    CHRONIC GLOMERULONEPHRITIS,
    NEPHROTOXIN DRUGS,
    INSECT BITE,
    IDIOPATHIC NEPHROTIC SYNDROME,
    SYSTEMIC LUPUS ERYTHEMATOSUS, OR
    DIABETES MELLITUS
MAY-BE-COMPLICATED-BY:
    HYPOVOLEMIA
    CELLULITIS
MAY-BE-CAUSE-OF:
    SODIUM RETENTION
DIFFERENTIAL-DIAGNOSIS:
    IF NECK VEINS ELEVATED,
        CONSIDER: CONSTRICTIVE PERICARDITIS
    IF ASCITES PRESENT,
        CONSIDER: CIRRHOSIS
    IF PULMONARY EMBOLI PRESENT,
        CONSIDER: RENAL VEIN THROMBOSIS

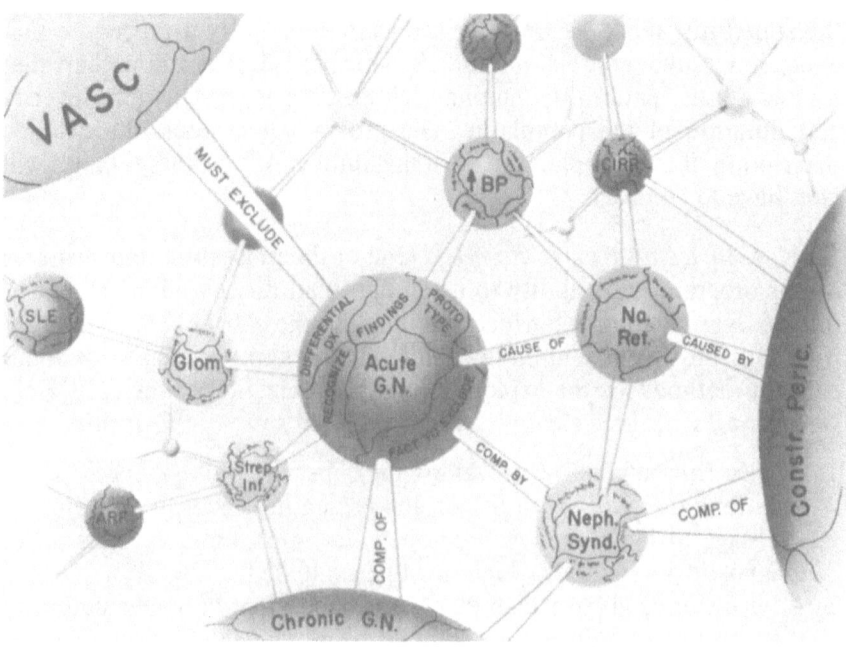

Figure 8. The associative (long-term) memory. The associative memory consists of a rich collection of knowledge about diseases, signs, symptoms, pathologic states, "real-world" situations, etc. Each point of entry into the memory allows access to many related concepts through a variety of associative links shown as rods. Each rod is labeled to indicate the kind of association it represents. Note that the dark gray spheres denote disease states, light gray spheres denote clinical states (e.g., nephrotic syndrome) and medium gray spheres denote physiologic states (e.g., sodium retention). Abbreviations used in this figure are Acute G.N. = acute glomerulonephritis; Chronic G.N. = chronic glomerulonephritis; VASC = vasculitis; CIRR = cirrhosis; Constr. Peric. = constrictive pericarditis; ARF = acute rheumatic fever; NA Ret. = sodium retention; SLE = systemic lupus erythematosus; ↑BP = acute hypertension; GLOM. = glomerulitis; Strep. inf. = streptococcal infection; Neph. Synd. = nephrotic syndrome.

◁ **Figure 7.** A typical "frame." Information about a disease, a physiologic state, etc., is stored in the form of a "frame" within the long-term memory. Included in the typical frame, as shown here for nephrotic syndrome, are descrptions of typical findings, numerical factors to be used in scoring, and links to other frames (e.g., "may-be-caused-by," "may-be-complicated-by"). There are also rules for excluding ("must-not-have") and satisfying ("is-sufficient") the fit of the frame to the case at hand. For further details, see text.

Characterizing Findings. After being presented with the chief complaint, the supervisor retrieves from the associative memory a procedure that characterizes that complaint in detail. This procedure is a flow chart that follows a "set" pattern in eliciting such features as the location, severity and duration of the complaint. The program later uses this detailed description of the complaint to limit the number of hypotheses that it will later have to consider.

Seeking Advice on How to Proceed. One of the most important features of our program is its ability to assemble small history-taking strategies into an over-all approach which is tailored to the case at hand. This ability is critically dependent on the availability of appropriate advice about efficient methods for the exploration and organization of the case. Here we shall present three examples of the program's use of this facility.

1. Advice can be given which alerts the supervisor to ask one or more questions that will "zero in" on the presenting problem and thus, at the stage of hypothesis generation (vida infra), limit the number of diagnostic possibilities which must be evaluated.
2. Advice can be given which guides the supervisor in its evaluation of the validity of information that is being presented. Such validity checks can be of several types. First, the program might point out that a finding itself is clearly in error, e.g., a weight gain of 50 pounds in 48 hours. Second, it might note that new information is inconsistent with other facts known about the patient, e.g., the presence of red cell casts in the absence of hematuria. Finally, it might indicate that a new finding contradicts a conclusion already drawn about the case.[5]
3. Advice can be given which alerts the supervisor to errors that might stem from a patient's misinterpretation of a particular sign or symptom. For example, if a patient complains of "blood in the urine," the supervisor is told that dark urine, which is attributed by the patient to blood, may be caused by the presence of bile, myoglobin or anthocyanins (from beets).

Hypothesis Generation. After the complaint has been characterized and all relevant advice has been acted upon, the supervisory program proceeds to generate working hypotheses. Hypothesis generation consists of moving frames from long-term memory to short-term memory where each frame plays a special role in guiding further exploration of the patient's problem.

Frames can exist in one of four states: dormant, semi-active, active and accepted. Initially, the short-term memory contains no frames; all

---

[5]The latter two kinds of advice would not be provided in the initial cycle which deals with the chief complaint because, at such an early stage, the short-term memory would not contain any detailed information about the patient.

frames are in the long-term memory and are said to be in the **dormant** state. In this nascent condition, however, some of the findings in the frames are associated with small, independent computer programs called **daemons.** A few of these daemons extend like tentacles from the frame into short-term memory (see Figure 9, "BEFORE"); these are primarily the daemons of those findings which are strongly suggestive of their associated frames. When the matching fact for a daemon is added to the short-term memory, the entire frame attached to the daemon is pulled into short-term memory (see Figure 9, "AFTER"). As pointed out, this process is synonymous with forming an hypothesis. Those frames that have entered short-term memory as hypotheses are called **active.**

As is reflected in Figure 9, "AFTER," frames "one link" away from an active frame are also affected in that during the activation process they are pulled closer to short-term memory. Consequently, more tentacles from such frames can reach into memory where they can now watch for their matching facts. These related frames, such as "Streptococcal infection" ("Strep. inf." in Figure 9, "AFTER"), are not allowed, however, to enter short-term memory. Moreover, their relatives, i.e., frames "two links removed" from the newly-active frame (e.g., "Acute Rheumatic Fever"), are not permitted to add more daemons on their own behalf. This "two-stage" limitation on hypothesis generation prevents an explosive expansion of the number of hypotheses that the program must consider at one time.

Those frames that have moved nearer to short-term memory and have added daemons to it are called **semiactive.** This state can be viewed as "sort of" thinking about something in the back of one's mind. If one of the daemons belonging to a semi-active frame finds a fact in short-term memory corresponding to its pattern, it, of course, causes the parent frame to be placed in short-term memory as an hypothesis and causes frames closely related to the new hypothesis to be pulled nearer to short-term memory.

Hypothesis Testing. Hypotheses generated by the program are evaluated to determine the extent to which they constitute reasonable explanations for the patient's condition. There are two aspects of this process. First, the fit of the case to the hypothesis (i.e., to a given frame) is appraised to determine whether the hypothesis can be accepted or rejected or whether more facts should be collected. Second, each hypothesis is examined to determine the extent to which it can account for all of the facts in the case.

The problem faced by the program in evaluating hypotheses is illustrated in Figure 10. In part A, we have represented schematically a perfect match between patient and disease prototype. An example of this situation would be a patient who has all the classic features of acute glomerulonephritis and no other abnormal findings. More typically, however,

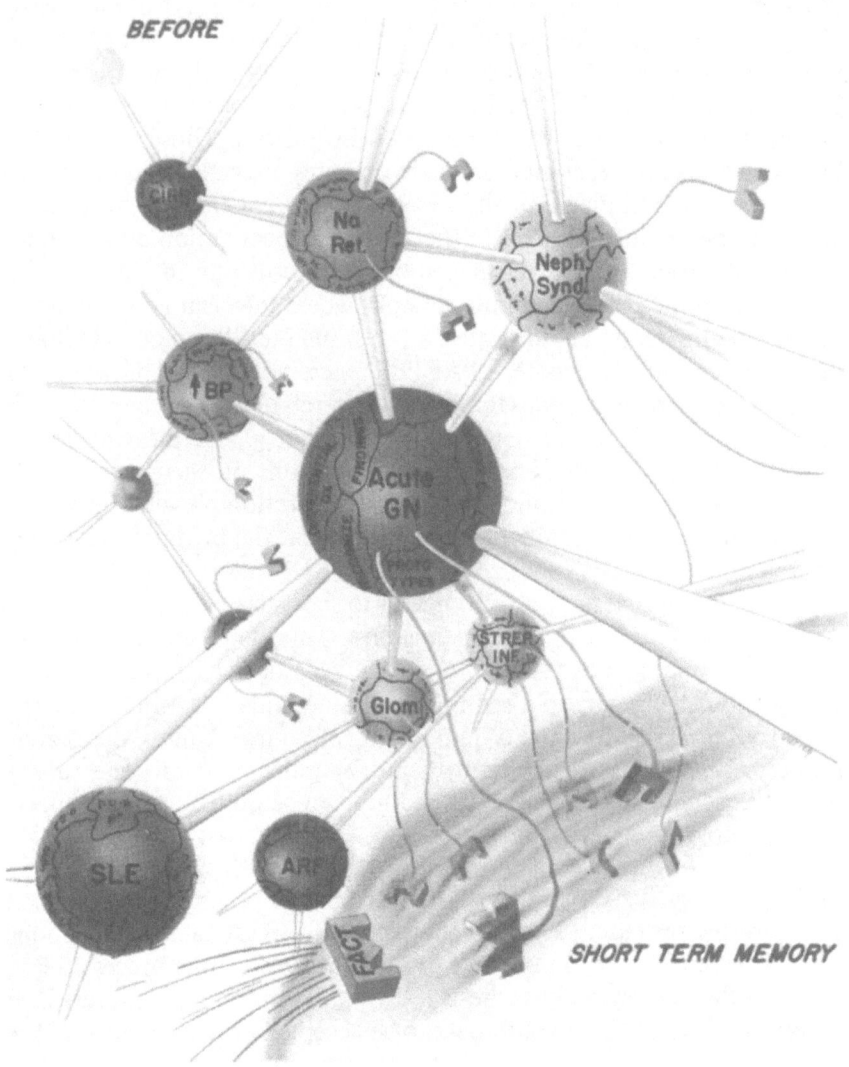

Figure 9. Hypothesis generation. **BEFORE:** in the nascent condition (when there are no hypotheses in short-term memory), tentacles (daemons) from some frames in long-term memory extend into the short-term memory where each constantly searches for a matching fact. **AFTER:** the matching of fact and daemon causes the movement of the full frame (in this case, acute glomerulonephritis) into short-

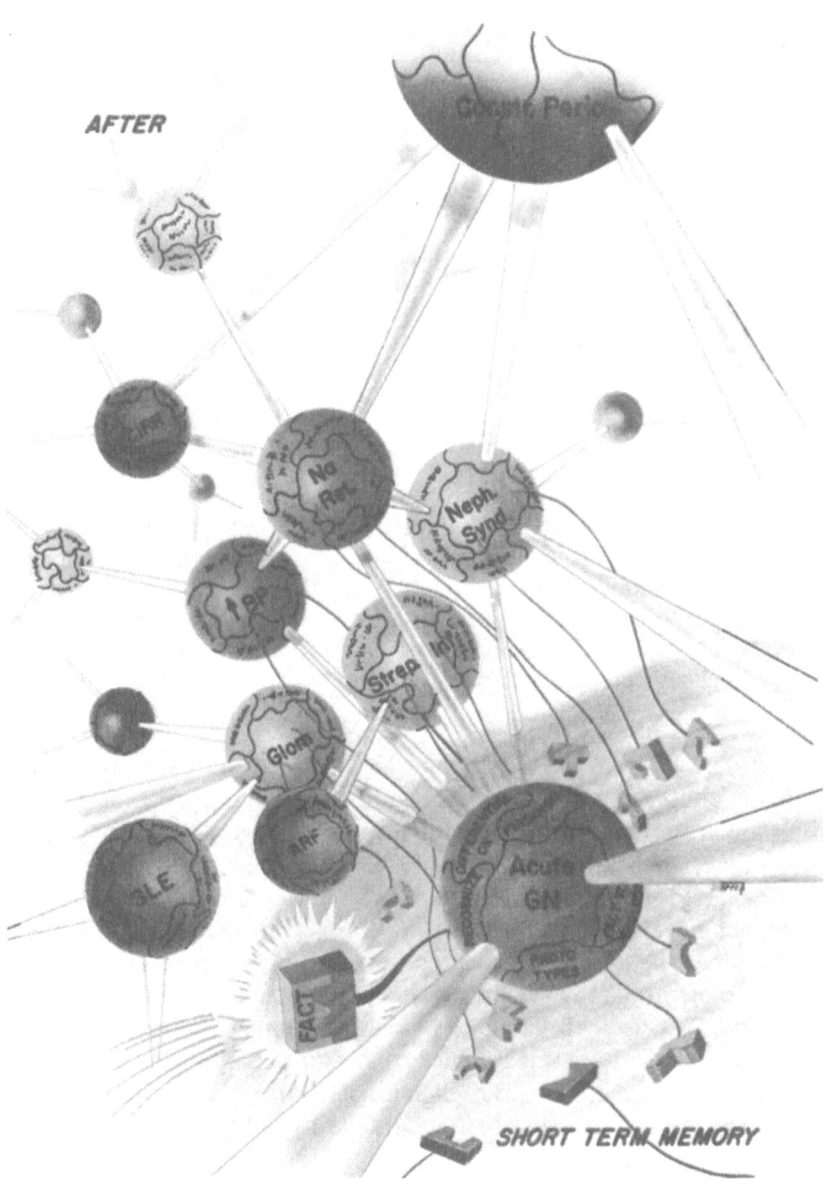

term memory. As a secondary effect, frames immediately adjacent to the activated frame move closer to short-term memory and are able to place additional daemons therein. Note that, to avoid complexity, the daemons on many of the frames are not shown. The abbreviations are the same as those used in Figure 8.

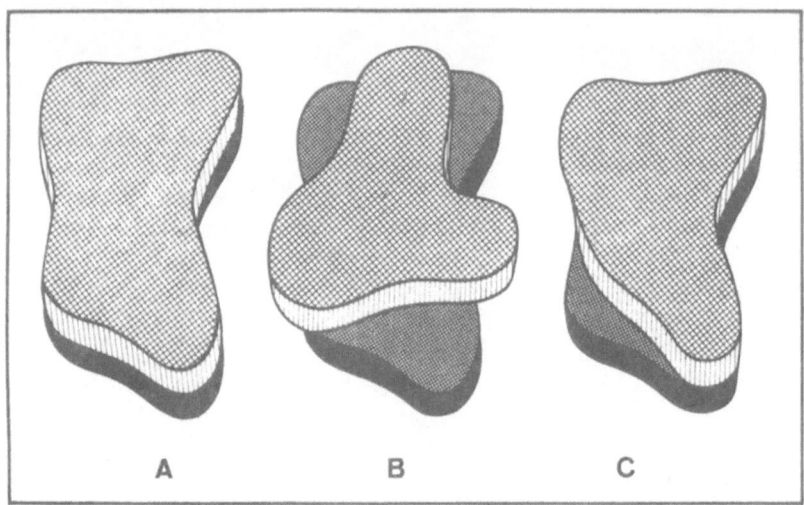

Figure 10. Schematic representation of pattern matching. Two wafers are shown in each instance, the lower one denoting the prototype being sought and the upper one denoting the case being tested. **CASE A:** an exact match. Every important feature of the prototype is found in the case and there are no features of the case which are not explained by the prototype. **CASE B:** there are featues of the case which are not explained by the prototype. **CASE C:** there are features in the prototype which are not found in the case.

findings are present in the patient which are not ordinarily seen in the state under consideration (part B, Figure 10), or findings characteristic of the state are missing from the patient (part C, Figure 10). The program uses numerical scores (to be discussed) to measure the degree of "fit" under each of these circumstances.

The fit of the case to the hypothesis serves to determine, as already mentioned, whether an active hypothesis can be accepted or rejected on the basis of the facts at hand or whether more information should be obtained. To help with this decision, each frame contains specific rules. For example, if idiopathic nephrotic syndrome is the hypothesis under consideration, and the program then learns that the patient has had gross hematuria, an **exclusionary rule** rejects the hypothesis and permanently removes the nephrotic syndrome frame from short-term memory. On the other hand, if the patient has both edema and massive proteinuria (protein excretion of greater than 5 g/24 hours), a **sufficiency rule** immediately accepts the nephrotic syndrome hypothesis.[6] In this accepted state, the hypothesis is asserted as if it were a fact. This new "fact" then is added

---

[6]Not only can disease frames be accepted or rejected, but frames corresponding to physiologic and clinical states can be similarly accepted or rejected.

to the short-term memory where it, in turn, can be found by daemons belonging to other frames. We should emphasize, however, that if later facts contradict the original conclusion, the acceptance is revoked.

In many instances, of course, there is no simple rule that can serve either to exclude or to establish a given hypothesis, and a **scoring process** is required. This scoring process uses numerical values (contained in the frame) which reflect the likelihood that various clinical findings will occur in the given disorder.[7] **Major** features are given more weight in the final scoring process than are **Minor** features.

Consider, for example, the nephrotic syndrome frame shown in Figure 7. Those features of the frame which can readily be identified as present or absent (e.g., low serum albumin concentration or heavy proteinuria) are given a numerical value. The remaining features are not initially assigned values; instead their contribution to the scoring process is determined by affiliated frames, e.g., hypovolemia is evaluated by means of a specific hypovolemia frame. Once such an affiliated frame has carried out its scoring function, the resulting value is passed on to the central frame. For example, acute glomerulonephritis receives a score for "elevated blood pressure in the absence of signs of chronic hypertension" from the acute hypertension frame to which it is related by a "may be complicated by" link. Similarly, the acute hypertension frame itself depends upon affiliated frames (e.g., hypertensive encephalopathy) for its own score. In some instances, such propagation of scoring proceeds through several levels.

If the score for an hypothesis exceeds a defined threshold, the frame is **accepted** by the supervisor. Similarly, if the score falls below a given threshold (i.e., if the hypothesis no longer fits the patient "well enough"), the supervisor forces the hypothesis into a semi-active state.

The ability of the hypothesis to account for the findings of the case is the extent to which all the facts of the case are explained by the hypothesis and its affiliated frames. The hypothesis of acute glomerulonephritis can explain, for example, both "low serum complement" and "oliguria"; the former finding is a part of the acute glomerulonephritis frame itself and the latter is a part of a closely linked frame, "acute renal failure." It cannot, however, account for the finding of long-standing hypertension. The program computes for each frame a value equal to the fraction of all findings in the patient profile which are explained by the hypothesis. This value and the measure of the fit of the case to the hypothesis are averaged and the hypotheses assigned a rank order based on the average.

Question Selection. After the supervisor has ranked the hypotheses, it seeks to gather more information about the patient in order to improve

---

[7]The weights associated with those features of the frame known to be present in the patient are summed and then normalized by the maximum attainable score.

its understanding of the clinical problem. The hypothesis that has received the highest over-all score is explored first, with the initial inquiries directed to the classic findings of the disorder. The answer to each question that is posed causes the re-evaluation of all hypotheses; as new information is obtained, the supervisor determines whether the leading hypothesis being pursued is still plausible, whether the hypothesis should now be accepted, or whether the hypothesis should be discarded from active consideration.

After the program has gathered information on prototypical features of the frame, it turns to questions about minor features of the disorder and then to inquiries about complications, etiologic factors and differential diagnoses. A change in the train of questioning usually indicates that, as the result of the continuous process of re-evaluation, a new hypothesis has moved into the leading position.[8]

The Cycle Repeated

Any new finding obtained in the course of the questioning process sets into motion a cycle which is the same as that just described for the chief complaint; each new sign or symptom is characterized, advice is sought, hypotheses are generated and tested, and additional questions are asked. In its cycle of response to new findings, the problem will, however, make use of the information acquired earlier (and hypotheses already generated) and will thus focus its questioning more sharply than if such a context were not available.

Controlling the Proliferation of Hypotheses

As discussed, the information-gathering and the diagnostic competence of the present illness program depend critically on its ability to quickly generate hypotheses to account for the patient's condition. For this reason, "aggressive" hypothesis-generation occurs even when only a few rather isolated facts are available. To avoid the excessive computational burden that would often be produced by such an "aggressive" strategy, the program employs several methods to restrict the number of hypotheses under active consideration.

Two of these methods have already been mentioned—the "zeroing-in" on a complaint and the "two-stage" process of hypothesis generation. A third method is the application of the **principle of parsimony.** Let us take,

---

[8]Sometimes the program considers a new hypothesis because of advice stored in the frame currently under consideration. For example, if nephrotic syndrome is the current hypothesis and symptoms suggestive of pulmonary emboli are reported, advice in the nephrotic syndrome frame will suggest that attention be shifted to renal vein thrombosis. The supervisor will then call up the questions designed to explore this latter possibility.

as an example, a patient with edema and massive proteinuria who is hypothesized to have nephrotic syndrome with sodium retention. If the program discovers that the patient has a positive test for anti-nuclear antibody, it does not simply add a new hypothesis, "systemic lypus erythematosus." Instead, it incorporates the hypothesis of "nephrotic syndrome with sodium retention" into a new, over-all hypothesis of "lupus erythematosus with nephrotic syndrome." If at a later time the more parsimonious hypothesis is rejected, the subhypotheses either are again given independent status as active frames or, alternatively, are returned to a dormant state.

Comments

The present report demonstrates that insights derived from the study of clinical cognition can be combined with advanced technics for computer simulation to create computer programs which possess a powerful problem-solving capability. The major technologic advance embodied in our program is the capacity to retrieve and apply knowledge **when that knowledge is required,** thus freeing the programmer from the virtually impossible task of specifying all contingencies in advance (as would be necessary, for example, in a branching flow-chart [1–3,10]). The key to implementation of the present system lies in the "goal-directed" nature of its operation. It is this goal-directed character which permits the supervisory program to select pertinent medical and real-world knowledge from the computer's memory and to dynamically assemble the many small problem-solving technics which efficiently guide the acquisition of additional clinical information. Another central feature of the program is the organization of its data base into an **associative memory** in which clusters of closely related facts about diseases and clinical states are stored in a fashion analogous to a richly cross-referenced encyclopedia. These groups of facts, called "frames" (Figure 7), are further organized into a *network* (Figure 8) which facilitates efficient retrieval of closely related blocks of information.

When the supervisory program is presented with a clinical problem (e.g., a chief complaint), it generates **hypotheses** about the case by moving frames from the **long-term** (associative) memory into **short-term** memory, where the frames interact with a profile of the patient's clinical data. When an hypothesis is generated, the supervisory program becomes "aware" of all the etiologies, complications and other features of the hypothesized condition because the frames describing such related facts are drawn into close proximity to short-term memory.

The goal of the program is, by evaluating these hypotheses, to arrive at the best possible diagnostic appraisal. To accomplish this purpose, the computer characterizes each finding in detail, seeks relevant advice from the associative memory and tests the hypotheses. The set of hypotheses under consideration also provides a framework within which additional information, both medical and real-world, is sought and interpreted. Throughout the questioning process, the supervisory program searches for inconsistencies in the information that it has obtained. When such inconsistencies arise, the program consults the memory for specific advice on how to deal with the conflicting information.

As new facts are obtained, additional hypotheses may be generated. From time to time, the supervisory program may also combine several hypotheses to form a more coherent diagnostic picture. As the questioning proceeds, all hypotheses under consideration are repetitively tested and "scored" to measure the "goodness of fit" between the description of the disease or physiologic state and the profile of facts about the patient. This testing provides the basis for either the acceptance or rejection of each hypothesis. At the termination of the questioning, the accepted hypotheses are listed and the other hypotheses are rank-ordered on the basis of the final score calculated for each.

The fundamental principles embodied in the present illness program are, we believe, applicable not solely to the problem of edema but are broadly relevant to the history-taking process. It is obvious, however, that many strategies other than those we have employed must be uncovered if a system such as we have described is to deal effectively with a wide range of clinical problems. In addition, numerous medical and real-world facts must be added to the program, and ingenious new technics must be devised that can deal with multiple coexisting diseases, that can draw appropriate inferences about the temporal aspects of a patient's history, and that can choose the point at which the questioning process should be terminated. Such problems obviously will require many years of intensive work for their solution.

The Problem of Scale

We believe that, over the long term, computer programs can be developed which should be capable not only of taking the present illness but also of assisting in virtually all aspects of patient management. If this view is correct, one must then ask whether the existing technology will be able to cope with the volume of information that might be required by such a system, i.e., will it be possible to store the requisite number of facts at a reasonable cost and to retrieve them in an efficient and effective manner? To answer this question we must first ask how much a computer program must "know" before it knows all of general internal medicine? Obviously, any calculation of this short must be highly speculative, but it seems certain that the program must have available at least that body of information which is contained in a standard textbook of medicine. As shown in section A of Table 1, it appears that each of the two most widely

Table 1. Estimate of Total Number of Facts Contained in Standard Textbooks of Medicine and in Representative Subspecialty Texts*

Title	Editors	Pages	Facts per Page (approx.)	Total Facts (approx.)
A. GENERAL INTERNAL MEDICINE				
Principles of Internal Medicine	Wintrobe [22]	2,035	100	200,000
Textbook of Internal Medicine	Beeson, McDermott [23]	1,892	100	190,000
B. SPECIALTY TEXTS				
Diseases of the Kidney	Strauss, Welt [24]	1,456	40	60,000
The Heart	Hurst [25]	1,755	50	90,000
Clinical Hematology	Wintrobe [26]	1,788	40	70,000

*Estimated as described under "Comments."

used textbooks of medicine contains on the order of 200,000 facts.[9] This estimate far understates, however, the total amount of information that is relevant to the practice of internal medicine. It is clear, for example, that there is a fund of basic science information used by the clinician which does not appear in such a textbook of medicine. To account for this body of data, we will double our estimate to a total of 400,000 facts. Finally, there is a considerable body of information about the real world (life insurance examinations, army physicals, time of day, seasons of the year) which, we will estimate, requires knowledge of still another 100,000 facts.[10] This brings us to a total of 500,000 facts. If we now double this value to take cognizance of possible underestimates, we arrive at an upper bound of approximately **one million facts** as the core body of information in general internal medicine.

The core knowledge embodied in the approximately ten separate subspecialties of internal medicine is, of course, considerably larger. To estimate the volume of clinical information basic to the entire domain of

[9]This estimate was arrived at by the crude technic of estimating the number of facts on several pages (not only basic facts, but the relationships between them) and multiplying this average by the total number of pages in the book. The major source of variability in such an estimate is the definition of what constitutes a *single fact*, because such a definition is to a certain extent arbitrary. In our calculations, we have used as a yardstick the amount of information that is treated as a single fact by our program; however, the choice of any other reasonable yardstick would not have changed our results appreciably.

[10]Note that we are only concerned with real-world knowledge that is relevant to medicine, not with all such knowledge possessed by the average person.

subspecialty medicine, we first have estimated the number of facts in text-books of nephrology, cardiology and hematology. As shown in section B of Table 1, each of these subspecialty treatises contains of the order of 60,000 facts. From this we estimate that the core body of information in all medical subspecialty texts combined is about 600,000 facts. If we assume that approximately one third of this information represents duplications among the specialty fields, we arrive at a total body of 400,000 facts, a value approximately twice that estimated for general internal medicine. Using the same ratio between facts and other kinds of relevant information that we have used in the case of general medicine, we calculate, correcting again for any possible underestimate, that the core of information in the subspecialties of internal medicine does not exceed **two million facts.**

Can the Knowledge-Base of Internal Medicine Be Stored at a Reasonable Cost?

Can two million facts be stored in a computer system at a reasonable cost? If we assume that each fact requires for its representation an average of 10 words of computer memory,[11] a computer storage capacity of 20 million words would be necessary. A memory of this size is certainly large and, if *core* storage were required, the cost would at present be prohibi-tive. Because only a small part of the data is used at any one time, an inexpensive mass storage device (such as a magnetic disc or drum) would be a practical alternative storage medium; even at present day prices, the cost would probably be no more than $20,000.

We should note, furthermore, that even with rapid progress in the development of "expert" consulting programs, it is unlikely that a system could reach the size we have envisioned in a period of less than 10 years. By that time, given the rapid evolution of computer technology, it is almost certain that a memory capable of storing two million facts could be purchased at a very low cost. From these considerations, it can rea-sonably be concluded that data storage will not be the limiting factor in the development of consulting capability within the computer. Indeed, even the storage of an additional large body of specialized information drawn from the medical literature, should, some years hence, pose no great technologic difficulties.

Can the Database of Internal Medicine Be Efficiently Managed?

The problems of organizing, retrieving and applying the *relevant* data are far more formidable than is the problem of data storage. We believe, how-

---

[11]We are considering this representation in a computer language such as LISP which is quite efficient at storing and retrieving the type of symbolic data that we envision will be used.

ever, that the task is probably not insuperable because in any given case only a very small fraction of the available knowledge needs to be retrieved. Furthermore, the retrieval of whatever information that is required will be greatly eased by the fact that pertinent information can be dealt with in the highly-organized clusters known as frames. Assuming that the average frame will contain on the order of 100 facts, only 20,000 frames would be required for the postulated data base of two million facts.

Probably the most difficult aspect of data management will be the problem of **coding,** the process of assuring that each fact is properly associated with other facts. Only if the large data base of internal medicine can be transferred automatically from English to the appropriate representation within the computer is there hope that serious errors, omissions and contradictions can be avoided. Current efforts to develop computer programs that understand English give promise that this fundamental problem will eventually be solved [21].

The arguments we have considered here have led us to the conclusion that, over the long-term, there are not likely to be intrinsic technologic constraints on the realization of a system capable of coping with all of internal medicine. In fact, the availability of increasingly powerful technology suggests a future in which computer programs may well "know" far more than any individual physician. For the short-term, however, we look towards the development of programs that know a great deal, but not all, of internal medicine.

Some Reflections on the Cognitive Process

As discussed earlier, the present illness simulation described here is based on insights derived from introspection and from observation of the problem-solving behavior of experienced clinicians. Here we offer a brief discussion of certain key ideas which we believe merit further study by investigators interested either in computer-aided decision-making or in clinical cognition.

Our study clearly illuminates an important difference between the expert in practice and the expert as often pictured in literature or folklore. The epitome of the expert in fiction is the detective who, through superior deductive powers and by sheer force of logic, organizes the facts at hand in a way that they lead to a single, inevitable conclusion. By contrast, the real-world clinician seems to rely much more heavily upon "guessing," his initial hypothesis typically being based on precious little data. These "guesses" are apparently prompted by patterns of clinical findings or by specific complaints which bring to mind particular diseases. The physician then tries to demonstrate the correctness of his "guesses," moving to new hypotheses only if his initial impressions prove untenable.

The rapidity with which the initial hypotheses are generated and the ostensibly fragile basis of the guessing process together constitute the most striking feature of the behavior of experienced clinicians. Often with only the age, sex and presenting complaint of the patient, the clinician unhesitatingly selects a single working hypothesis. Even in ambiguous situations, he rarely begins with more than a few hypotheses.

Another characteristic of the experienced physician is the fashion in which he continually pares his list of diagnostic possibilities. He discards some, accepts others and often combines individual possibilities into a single, new, integrated hypothesis. In this way, he is generally able to limit sharply the number of diagnoses which he must actively consider. We can understand the value of such a sharp focus when we consider that, in taking a present illness, the physician can gather only a small fraction of the potential set of facts concerning the patient and must therefore seek information very selectively. In consequence, he must find a context within which to properly focus his questioning and to organize the information that he obtains.

Because the initial hypotheses are usually generated on the basis of relatively few facts, they will often later prove to be incorrect. In such cases, how does the experienced clinician proceed to undo any "damage" done by his aggressive hypothesis generation? Our observations suggest that he often employs the rather efficient strategy of associating one hypothesis with others with which it may be readily confused (e.g., "multiple pulmonary emboli are often confused with cardiomyopathy"). By explicitly remembering such situations, the physician can move directly from an hypothesis which has become suspect to one which offers another plausible explanation for the presenting findings.

Unlike the seasoned clinician, the medical student or young physician does not have an extensive knowledge of such relations and so is unlikely to move from one hypothesis to another in such a skillful fashion. Therefore, the novice who acts aggressively in hypothesis generation risks making serious errors. We have observed that the student or house officer, apparently to counter this problem, often approachs the diagnostic process in a highly structured, methodical fashion. Similarly we have noted that the experienced physician performing outside his area of expertise uses a far more structured approach than is his usual custom. The seasoned clinician's expertise in taking a present illness thus appears to derive in considerable part from a complex set of associations and from a familiarity with many alternative scenarios within each of his "frames."

We believe that the experimental methods utilized in the present study, if extensively employed, will provide important new insights into the process of clinical problem-solving. Furthermore, as our understanding of problem-solving processes grows, it seems likely that the study of clinical cognition will assume a significant place in the medical curriculum. Such increased attention to this neglected aspect of medical education should eventually make an important contribution to improving the quality of physician performance.

Acknowledgment. This research was supported in part by the Health Resources Administration, U.S. Public Health Service, under Grant 1 R01 MB 00107-01 from the Bureau of Health Manpower and under Grant HS 00911-01 from the National Center for Health Services Research.

References

[1] Slack WV, Hicks GP, Reed CE, VanCura LJ: A computer-based medical history system. *N Engl J Med* 274:194, 1966.

[2] Bleich HL: Computer evaluation of acid-base disorders. *J Clin Invest* 48: 1689, 1969.

[3] Bleich HL: Computer-based consultation, electrolyte and acid-base disorders. *Am J Med* 53:285, 1972.

[4] Peck CC, Scheiner LB, Martin CM, Combs MD, Melmon KL: Computer-assisted digoxin therapy. *N Engl J Med* 289: 441, 1973.

[5] Jelliffe RW, Buell J, Kalaba R: Reduction of digitalis toxicity by computer-assisted glycoside dosage regimens. *Ann Intern Med* 77: 891, 1972.

[6] Gorry GA, Kassirer JP, Essig A, Schwartz WB: Decision analysis as the basis for computer-aided management of acute renal failure. *Am J Med* 55:473, 1973.

[7] Shortliffe EH, Axline SG, Buchanan BG, Merigan TC, Cohen SN: An artificial intelligence program to advise physicians regarding antimicrobial therapy. *Comput Biomed Res* 6:544, 1973.

[8] Kulikowski CA, Weiss S, Safir A: Glaucoma diagnosis and therapy by computer. Proceedings of Annual Meeting of Association for Research in Vision and Ophthalmology, Sarasota, Florida, May 1973.

[9] Pople H, Werner G: An information processing approach to theory formation in biomedical research. Proceedings of the AFIPS Spring Joint Computer Conference, 1972, p 1125.

[10] Stead WW, Heyman A, Thompson HK, Hammond WE: Computer-assisted interview of patients with functional headache. *Arch Intern Med* 129:950, 1972.

[11] Winston PH: New Progress in Artificial Intelligence TR-310, Cambridge, Massachusetts, The Artificial Intelligence Laboratory, Massachusetts Institute of Technology, 1974.

[12] Feigenbaum EA, Feldman J, eds: Computers and Thought. New York, McGraw-Hill, 1963. This book is a collection of papers on a variety of artificial intelligence problems: chess playing, solving geometry problems, theorem proving, and some cognitive simulation experiments. Written in the late 1950's and early 1960's, these papers reflect the then current emphasis on highly-structured problem domains and more or less formal problem-solving methods. Of particular interest is Minsky's paper "Steps Toward Artificial Intelligence," a lucid discussion of the state of the art at that time.

[13] Bell DA: Intelligent Machines (An introduction to cybernetics). London, Pitman, 1962.

[14] McCarthy J: Programs with common sense, Semantic Information Processing (Minsky M, ed), Cambridge, MIT Press, 1968.

[15] Shannon CE, Weaver W: The Mathematical Theory of Communication, Urbana, Ill., University of Illinois Press, 1949.

[16] Minsky M: A framework for representing knowledge. The Psychology of Computer Vision (Winston PH, ed), New York, McGraw-Hill, 1975.

[17] Minsky M (ed): Semantic Information Processing. Cambridge, MIT Press, 1968. This collection of papers considers such problems as finding analogies between things, making logical and non-logical inferences, and engaging in a coherent dialogue with a person. The papers are interesting in their own right but also by virtue of the historical perspective which they provide on the shift toward the intensive study of knowledge by artifical intelligence researchers.

[18] Newell A, Simon HA: Human Problem Solving, Englewood Cliffs, New Jersey, Prentice-Hall, 1972. This book is a comprehensive summary of work done over the past 15 years by two of the most prolific and important workers in the field. Through the extensive use of protocol analysis and computer simulation, Newell and Simon have developed a theory of human problem solving which has had a widespread influence in artificial intelligence, psychology and management science. This book is of interest as a review of their work, as a statement of their theory and as an exposition of a powerful methodology for the study of cognition.

[19] Sussman GJ, McDermott DV: From PLANNER to CONNIVER—a genetic approach. Proceedings of the Fall Joint Computer Conference, Anaheim, Calif., 1972, p 1171.

[20] Feinstein AR: Clinical Judgment, Baltimore, Williams & Wilkins, 1967.

[21] Shank RC, Colby KM (eds): Computer Models of Thought and Language, San Francisco, WH Freeman & Co., 1973. This collection of papers describes more recent attempts to model human psychological processes on a computer. On the whole, the work reported in this book concerns the creation and testing of more complex models of human behavior than those discussed in the books described above. Also of interest are the contributions of psychologists to the volume. Their papers represent a slight, but significant trend toward the casting of psychological theories in computer programs rather than in the more conventional literary or mathematical forms.

[22] Wintrobe MM (ed): Harrison's Principles of Internal Medicine, New York, McGraw-Hill, 1974.

[23] Beeson PW, McDermott W (eds): Cecil-Loeb Textbook of Medicine, Philadelphia, W.B. Saunders, 1975.

[24] Strauss MB, Welt LG: Diseases of the Kidney, Boston, Little, Brown & Co., 1971.

[25] Hurst JW: The Heart, Arteries and Veins, New York, McGraw-Hill, 1974.

[26] Wintrobe MM: Clinical Hematology, Philadelphia, Lea & Febiger, 1974.

INTERNIST-I, An Experimental Computer-Based Diagnostic Consultant for General Internal Medicine

Randolph A. Miller, Harry E. Pople, Jr., and Jack D. Myers

INTERNIST-I is an experimental computer program capable of making multiple and complex diagnoses in internal medicine. It differs from most other programs for computer-assisted diagnosis in the generality of its approach and the size and diversity of its knowledge base. To document the strengths and weaknesses of the program we performed a systematic evaluation of the capabilities of INTERNIST-I. Its performance on a series of 19 clinicopathological exercises (Case Records of the Massachusetts General Hospital) published in the *Journal* appeared qualitatively similar to that of the hospital clinicians but inferior to that of the case discussants. The evaluation demonstrated that the present form of the program is not sufficiently reliable for clinical applications. Specific deficiencies that must be overcome include the program's inability to reason anatomically or temporally, its inability to construct differential diagnoses spanning multiple problem areas, its occasional attribution of findings to improper causes, and its inability to explain its "thinking." (N Engl J Med. 1982; 307:468–76.)

INTERNIST-I, an experimental program for computer-assisted diagnosis in general internal medicine, differs considerably in scope from other medical diagnostic computer programs. In the past, techniques including mathematical modeling, use of Bayesian statistics, pattern recognition, and other approaches [1–3] have been shown to be useful in circumscribed areas, such as the differential diagnosis of abdominal pain [4] and the diagnosis and treatment of meningitis [5]. However, no program developed for use in a limited domain has been successfully adapted for more generalized use. From its inception, INTERNIST-I has addressed the problem of diagnosis within the broad context of general internal medicine [6–8]. Given a patient's initial history, results of a physical examination, or laboratory findings, INTERNIST-I was designed to aid the physician with the patient's workup in order to make multiple and

Reprinted with permission from the *New England Journal of Medicine, 307,* pp. 468–476, 1982.

complex diagnoses. The capabilities of the system derive from its extensive knowledge base and from heuristic computer programs that can construct and resolve differential diagnoses.

The INTERNIST-I program represents an example of applied symbolic reasoning ("artificial intelligence"). A variety of such techniques have been developed by computer scientists in an attempt to model the thought processes and problem-solving methods employed by human beings [9,10]. An important aspect of the INTERNIST-I approach to computer-assisted diagnosis is that the program attempts to form an appropriate differential diagnosis in individual "problem areas." A problem area is defined as a selected group of observed findings, the differential diagnosis of which forms what is assumed to be a mutually exclusive, closed (i.e., exhaustive) set of diagnoses. Physicians routinely construct such closed differential diagnoses on the basis of causal considerations (e.g., bacterial pneumonias) or pathoanatomic considerations (e.g., causes of obstructive jaundice). By constructing specific differential diagnoses to address identified problem areas, a physician or computer program can narrow the set of possible diagnoses from all known diseases to well-defined collections of competing diagnoses in a small number of categories. Heuristic principles, such as diagnosis by exclusion, can then be employed to resolve each differential diagnosis. The use of such strategies in INTERNIST-I represents an attempt to model the behavior of physicians.

Reported below is the first systematic evaluation of INTERNIST-I. The purpose of the study was to illustrate the strengths and weaknesses of the program and to provide a rough estimate of its clinical acumen. The trial was conducted with clinicopathological conferences (CPCs) that had been published in the *Journal* but had not previously been analyzed by the system. The CPCs fulfill the criteria of being diagnostically challenging cases and of containing sufficiently detailed information to allow computer analysis. The evaluation was not intended to validate INTERNIST-I for clinical use. CPCs should not be used for such a purpose, and as the trial demonstrated, the program does not yet possess sufficient reliability for clinical application. Nevertheless, INTERNIST-I performed remarkably well, considering the simple, ad hoc nature of its algorithms.

The INTERNIST-I Data Base

A medical knowledge base must meet the needs of any associated diagnostic programs. In particular, the INTERNIST-I knowledge base was designed to permit the consultant program to construct and resolve differential diagnoses. The knowledge base incorporates individual disease profiles, which list findings that can occur in patients with each illness.

By inverting the disease profiles with use of a computer program, an exhaustive differential diagnosis for each finding is obtained; these manifestation-based differential-diagnosis lists are retained as part of the knowledge base. The diagnostic program can use these lists to construct differential diagnoses in clinical cases.

How to group potential diagnoses into relevant problem areas is a separate consideration. The individual diseases in the INTERNIST-I knowledge base are part of a "disease hierarchy" that is organized from the general to the specific. For example, acute viral hepatitis is classified as a hepatocellular infection, hepatocellular infection is a subclass of diffuse hepatic parenchymal disease, and diffuse hepatic parenchymal disease falls into the category of hepatic parenchymal disease, which is a major subclass of diseases of the hepatobiliary system. Initially it was thought that access to the disease hierarchy would allow INTERNIST-I to construct appropriate differential diagnoses (i.e., problem areas) based on higher-level concepts such as hepatocellular infection. If several diagnoses representing types of hepatocellular infection were under consideration, it would be simple to create a problem area for hepatocellular infection. However, early experience with the system showed that a rigid hierarchical classification scheme was inadequate, since a single disease often merits simultaneous categorization under more than one heading. Infectious mononucleosis is both a hepatocellular infection and a type of infectious lymphadenopathy. Hierarchical classification would require that it be listed as one or the other, but not both. An additional concern is that diseases that present in a fixed manner in any one person may present differently in different patients. For example, alcoholic hepatitis may occur with a predominance of intrahepatic cholestasis in one patient and with massive hepatocellular necrosis in another. Solution of the classification problem entailed the development of algorithms (discussed below) that permit INTERNIST-I to construct problem areas in an ad hoc manner.

The building block for the INTERNIST-I knowledge base is the individual disease. For each diagnosis entered into the system, a disease profile is constructed. The disease profile consists of findings (historical items, symptoms, physical signs, and laboratory abnormalities) that have been reported to occur in association with the disease, including demographic data and predisposing factors. Two clinical variables are associated with each manifestation in an INTERNIST-I disease profile: an evoking strength and a frequency. The evoking strength answers the question, Given a patient with this finding, how strongly should I consider this diagnosis to be its explanation? The frequency is an estimate of how often patients with the disease have the finding. In addition, each manifestation is assigned a disease-independent import. The import is the global importance of the manifestation—that is, the extent to which one is compelled to explain its presence in any patient. Although the evoking

Table 1. Interpretation of Evoking Strengths

Evoking Strength	Interpretation
0	Nonspecific—manifestation occurs too commonly to be used to construct a differential diagnosis
1	Diagnosis is a rare or unusual cause of listed manifestation
2	Diagnosis causes a substantial minority of instances of listed manifestation
3	Diagnosis is the most common but not the overwhelming cause of listed manifestation
4	Diagnosis is the overwhelming cause of listed manifestation
5	Listed manifestation is pathognomonic for the diagnosis

strengths, frequencies, and imports are expressed as numbers (on a scale of 0 to 5 or 1 to 5) in the INTERNIST-I knowledge base, it is important to remember that they represent a shorthand for judgmental information, as their suggested interpretations in Tables 1 through 3 indicate. True quantitative information does not exist in the medical literature in most cases; the numbers used by INTERNIST-I are judgmental in that they are compiled after a review of the available knowledge.

The current INTERNIST-I knowledge base, which represents 15 person-years of work, encompasses over 500 individual disease profiles (an example appears in Table 4) and approximately 3550 manifestations of disease. The disease profiles have been generated by a review of the literature and by consultation with expert clinicians. In addition to the disease profiles, the knowledge base details relations among diagnoses and among manifestations. Within INTERNIST-I, important high-level pathophysiologic states (such as acute left ventricular failure, chronic congestive left heart failure, prerenal azotemia, and chronic uremia) are

Table 2. Interpretation of Frequency Values

Frequency	Interpretation
1	Listed manifestation occurs rarely in the disease
2	Listed manifestation occurs in a substantial minority of cases of the disease
3	Listed manifestation occurs in roughly half the cases
4	Listed manifestation occurs in the substantial majority of cases
5	Listed manifestation occurs in essentially all cases—i.e., it is a prerequisite for the diagnosis

Table 3. Interpretation of Import Values

Import	Interpretation
1	Manifestation is usually unimportant, occurs commonly in normal persons, and is easily disregarded
2	Manifestation may be of importance, but can often be ignored; context is important
3	Manifestation is of medium importance, but may be an unreliable indicator of any specific disease
4	Manifestation is of high importance and can only rarely be disregarded as, for example, a false-positive result
5	Manifestation absolutely must be explained by one of the final diagnoses

profiled as if they were diseases. The knowledge base contains links between such "diseases" and other diseases. The links are used to express causality or a predisposition of patients with one disease to have another. Because INTERNIST-I formulates and resolves problem areas serially, it can piece together interdependent components of a multisystem illness one by one, using the links in the data base to promote consideration of diseases related to previously concluded diagnoses. The total number of links among the 500 diagnoses in the data base is about 2600. The 3550 manifestations in the INTERNIST-I knowledge base are not independent. Men do not have oligomenorrhea, and a patient with oligomenorrhea must be presumed to be female. The knowledge base includes the properties of each manifestation that specify how its presence or absence may influence the presence or absence of other manifestations. There are about 6500 such relations detailed in the knowledge base.

The Diagnostic Algorithms

The problem-solving algorithms represent the intellectual core of the INTERNIST-I system. Although the scoring mechanism described below manipulates probabilistic data (evoking strengths, frequencies, and imports), it must be emphasized that the behavior of INTERNIST-I results primarily from the application of two heuristic principles: the formation of problem areas through a partitioning algorithm and the conclusion of diagnoses within problem areas, using strategies such as diagnosis by exclusion.

The following steps are taken during an INTERNIST-I diagnostic consultation. (Please refer to the Appendix for an annotated sample case analysis taken from a CPC published in the *Journal* [11].)

(1) Initial positive (present) and negative (absent) findings in the patient are entered by the user. As each new positive manifestation is

Table 4. A Sample Manifestations List*

DISPLAY WHICH MANIFESTATION LIST?
ALCOHOLIC HEPATITIS

AGE 16 TO 25 . . . 0 1
AGE 26 TO 55 . . . 0 3
AGE GTR THAN 55 . . . 0 2
ALCOHOL INGESTION RECENT HX . . . 2 4
ALCOHOLISM CHRONIC HX . . . 2 4
SEX FEMALE . . . 0 2
SEX MALE . . . 0 4
URINE DARK HX . . . 1 3
WEIGHT LOSS GTR THAN 10 PERCENT . . . 0 3
ABDOMEN PAIN ACUTE . . . 1 2
ABDOMEN PAIN COLICKY . . . 1 1
ABDOMEN PAIN EPIGASTRIUM . . . 1 2
ABDOMEN PAIN NON COLICKY . . . 1 2
ABDOMEN PAIN RIGHT UPPER QUADRANT . . . 1 3
ANOREXIA . . . 0 4
DIARRHEA ACUTE . . . 1 2
MYALGIA . . . 0 3
VOMITING RECENT . . . 0 4
ABDOMEN BRUIT CONTINUOUS RIGHT UPPER QUADRANT . . . 1 2
ABDOMEN BRUIT SYSTOLIC RIGHT UPPER QUADRANT . . . 1 2
ABDOMEN TENDERNESS RIGHT UPPER QUADRANT . . . 2 4
CONJUNCTIVA AND/OR MOUTH PALLOR . . . 1 2
FECES LIGHT COLORED . . . 1 2
FEVER . . . 0 4
HAND(S) DUPUYTRENS CONTRACTURE(S) . . . 1 2
JAUNDICE . . . 1 3
LEG(S) EDEMA BILATERAL SLIGHT OR MODERATE . . . 1 2
LIVER ENLARGED MASSIVE . . . 1 2
LIVER ENLARGED MODERATE . . . 1 3
LIVER ENLARGED SLIGHT . . . 1 2
PAROTID GLAND(S) ENLARGED . . . 1 2
SKIN PALLOR GENERALIZED . . . 0 2
SKIN PALMAR ERYTHEMA . . . 1 3
SKIN SPIDER ANGIOMATA . . . 2 3
SKIN TELANGIECTASIA . . . 1 1
ALKLAINE PHOSPHATASE BLOOD GTR THAN 2 TIMES
    NORMAL . . . 1 2
ALKALINE PHOSPHATASE BLOOD INCREASED NOT OVER 2 TIMES
    NORMAL . . . 1 4
BILIRUBIN BLOOD CONJUGATED INCREASED . . . 2 4
BILIRUBIN URINE PRESENT . . . 2 4
CHOLESTEROL BLOOD DECREASED . . . 2 2
CHOLESTEROL BLOOD INCREASED . . . 1 2
HEMATOCRIT BLOOD LESS THAN 35 . . . 1 3
HEMOGLOBIN BLOOD LESS THAN 12 . . . 1 3

Table 4. Continued.

KETONURIA ... 1 2
PROTEINURIA ... 1 2
SGOT 120 TO 400 ... 2 3
SGOT 40 TO 119 ... 2 3
SGOT GTR THAN 400 ... 1 2
UREA NITROGEN BLOOD LESS THAN 8 ... 2 2
UROBILINOGEN URINE ABSENT ... 1 1
UROBILINOGEN URINE INCREASED ... 2 4
WBC 14000 TO 30000 ... 0 3
WBC 4000 TO 13900 PERCENT NEUTROPHIL(S) INCREASED ... 0 3
WBC LESS THAN 4000 ... 1 1
ACTIVATED PARTIAL THROMBOPLASTIN TIME INCREASED ... 1 3
ANTIBODY MITOCHONDRIAL ... 1 1
ANTIBODY SMOOTH MUSCLE ... 2 3
BSP RETENTION INCREASED ... 1 5
ELECTROPHORESIS SERUM ALBUMIN DECREASED ... 2 4
ELECTROPHORESIS SERUM GAMMA GLOBULIN INCREASED ... 2 4
FACTOR VII PROCONVERTIN DECREASED ... 1 2
LDH BLOOD INCREASED ... 1 3
MAGNESIUM BLOOD DECREASED ... 2 2
PROTHROMBIN TIME INCREASED ... 2 3
SGPT 200 TO 600 ... 1 2
SGPT 40 TO 199 ... 2 3
SGPT GTR THAN 600 ... 1 1
LIVER BIOPSY BILE PLUGGING ... 1 2
LIVER BIOPSY FATTY METAMORPHOSIS ... 2 4
LIVER BIOPSY FOCAL NECROSIS AND INFLAMMATION ... 2 5
LIVER BIOPSY HEPATOCELLULAR NECROSIS MARKED ... 2 3
LIVER BIOPSY MALLORY BODIES ... 3 3
LIVER BIOPSY PERIPORTAL FIBROSIS MILD ... 1 3
LIVER BIOPSY PERIPORTAL INFILTRATION NEUTROPHIL(S) ... 3 5
LIVER BIOPSY PERIPORTAL INFILTRATION ROUND CELL(S) ... 1 2
LIVER BIOPSY SMALL BILE DUCT(S) PROMINENT ... 1 2

LINKS FOR ALCOHOLIC HEPATITIS:

Predisposes to	MALLORY WEISS SYNDROME ... 1 1
Causes	SINUSOIDAL OR POSTSINUSOIDAL PORTAL HYPERTENSION ... 1 2
Causes	HEPATIC ENCEPHALOPATHY ... 2 2
Causes	RENAL FAILURE SECONDARY TO LIVER DISEASE ⟨HEPATORENAL SYNDROME⟩ ... 2 2
Coincident with	PANCREATITIS ACUTE ... 2 2
Precedes	MICRONODAL CIRRHOSIS ⟨LAENNECS⟩ ... 2 3

*The first number after each manifestation is its evoking strength for the diagnosis; the second is the frequency of the manifestation in the disease.

encountered, the program retrieves its complete differential diagnosis from the inverted disease profiles in the knowledge base. A "disease hypothesis" is created for each item on the manifestation's differential-diagnosis list. A master list of all such disease hypotheses is maintained. Higher-level concepts from the classification hierarchy are retained on the differential-diagnosis list as long as the diagnoses that they subsume are indistinguishable in their ability to explain the observed data. The master differential list therefore comprises all possible diagnoses that can explain any of the observed findings (taken either individually or in groups).

(2) For each disease hypothesis, four lists are maintained: all positive manifestations in the patient that are explained by the disease hypothesis (i.e., findings matching the disease profile stored in the data base); all manifestations that might occur in a patient with the disease but are known to be absent in the patient being considered; all manifestations that are present in the patient but are not explained by the disease hypothesis, i.e., not found on the disease profile (these manifestations represent either "red herrings" or items that would have to be explained by a second disease present in the patient); and manifestations that are on the disease's profile but about which nothing is known (this list is used in determining which questions to ask).

(3) Each hypothesis on the master list of diagnoses is given a score. Scores are calculated as the sum of a positive and a negative component as follows. The positive component includes the weights of all manifestations explained by the hypothesis, based on the evoking strengths of the observed manifestations for the diagnosis. A nonlinear weighting scheme is used: An evoking strength of 0 counts as 1 point, a strength of 1 counts as 4 points, a 2 counts as 10 points, a 3 counts as 20, a 4 as 40, and a 5 as 80. Any disease hypothesis related to a previously concluded diagnosis (through links in the data base) is given a bonus score. The bonus awarded is 20 points times the frequency number listed for the hypothesized diagnosis in the disease profile of the concluded diagnosis.

The negative component includes the weight of all manifestations that are expected to occur in patients with the disease but are absent in the patient under consideration. A nonlinear scale based on the expected frequency of the manifestation in the disease is used: A frequency of 1 counts as -1 point, a 2 as -4 points, a 3 as -7 points, a 4 as -15 points, and a 5 as -30 points. Also included are the weights of all manifestations present in the patient but not explained by the hypothesized diagnosis. The import (clinical importance) of each manifestation is used to assess this penalty: An import of 1 counts as -2 points, a 2 as -6 points, a 3 as -10 points, a 4 as -20 points, and a 5 as -40 points. The net score for any disease hypothesis is thus the sum of the above four component weights.

(4) After all disease hypotheses have been scored, the master list of all

hypotheses is sorted by descending score. Diagnoses whose scores fall a threshold number of points below the topmost diagnosis are temporarily discarded as unattractive. They may be reconsidered, however, if further evidence obtained during the case analysis raises their scores above the threshold (relative to the topmost diagnosis).

(5) At this point, the sorted master differential-diagnosis list is a heterogeneous grouping of many disease hypotheses. A critical step in the diagnostic logic of INTERNIST-I is to delineate a set of competitors for the topmost diagnosis (i.e., to create a problem area containing the topmost disease hypothesis). Only one of the set of diseases in a properly defined problem area is likely to be present in a patient. Problem-area construction is carried out by the INTERNIST-I partitioner, which employs a remarkably powerful yet simple heuristic rule. The rule states, "Two diseases are competitors if the items not explained by one disease are a subset of the items not explained by the other; otherwise, they are alternatives (and may possibly coexist in the patient)." To paraphrase, if Disease A and Disease B, taken together, explain no more observed manifestations than either one does when taken alone, the diseases are classified as competitors. Competitors for the likeliest diagnosis are identified from the master differential list by the partitioning rule; including the topmost diagnosis, they constitute the "current problem area." Because INTERNIST-I defines problem areas in this ad hoc manner, its differential diagnoses will not always resemble those constructed by clinicians.

(6) Once the problem area containing the most attractive diagnosis has been selected, criteria for establishing a definitive diagnosis can be applied. If the problem area contains only the topmost diagnosis, INTERNIST-I will immediately decide on (conclude) that diagnosis. If there is more than one diagnosis in the problem area, INTERNIST-I directly concludes the leading diagnosis when its score is 90 or more points higher than the nearest competitor. The value of 90 was chosen because it slightly exceeds the weight carried by a pathognomonic finding (80 points). This method of concluding a diagnosis is a hallmark of INTERNIST-I. The absolute score of the diagnosis does not matter. The only point of importance is whether the diagnosis is sufficiently higher in score than its reasonable competitors (other diagnoses that explain the same set of findings).

(7) If it is not possible to conclude a diagnosis (which by default means that the current problem area contains more than one hypothesis), one of three questioning strategies is selected: "pursuing," "ruling out," and "discriminating." The pursuing mode is selected if the second best contender is 46 to 89 points behind the topmost diagnosis. In the pursuing mode, questions are asked to establish the topmost diagnosis, since it is close to fulfilling the criteria for conclusion. The questions asked are those that are most specific for the leading diagnosis (i.e., those with high evoking strengths). If there are five or more diagnoses within 45 points

of the topmost diagnosis, the ruling-out mode is used. Questions that have high frequency numbers under the contenders are asked, with the expectation that several negative responses will remove some of the diagnoses from contention. The discriminating mode is used when there are two to four diagnoses within 45 points of the leading diagnosis. The questions asked attempt to maximize the spread in scores.

(8) In order to improve the efficiency of computations, questions are asked in small groups. The level of questioning is escalated (from history to physical-examination findings to gradations of laboratory results) only after the useful questions in a previous category have been exhausted. After the answers are processed, the disease hypotheses are again scored and partitioned. A new differential diagnosis is formed on the basis of the (possibly) new topmost diagnosis. This ad hoc method for constructing a differential diagnosis gives INTERNIST-I seemingly intelligent behavior, since the program will often change focus from one problem area to another when questioning in the first area has been counterproductive.

(9) When a diagnosis is concluded, all observed manifestations explained by the diagnosis are removed from future consideration. The program then recycles, using the remaining unexplained positive findings. Subsequently findings are marked as explained when a previously concluded diagnosis can account for them. However, it is not possible to undo a previous diagnostic conclusion when contradictory evidence becomes available.

(10) When a problem area contains more than one disease hypothesis and all useful lines of questioning have been exhausted (without meeting the criteria for concluding the topmost diagnosis), the program will defer making a diagnosis in that problem area. Diagnoses in the problem area are then displayed by descending score, along with an explanation that the differential diagnosis cannot be resolved.

(11) When all remaining manifestations have an import of 2 or less, the program stops.

An Evaluation of INTERNIST-I

We have completed a preliminary evaluation of INTERNIST-I. The program was evaluated to compare its clinical acumen with that of human experts and to highlight its strengths and weaknesses. CPCs published in the *Journal* as Case Records of the Massachusetts General Hospital were used for the computer analysis. During the trial, only the published findings available to the case discussant were presented to INTERNIST-I (i.e., only findings mentioned before the presentation of the pathological findings). The knowledge base of INTERNIST-I was not altered during the course of the evaluation.

During the development of INTERNIST-I, hundreds of miscellaneous cases, both simple and complex, have been presented to the system in order to evaluate and improve the knowledge base and the diagnostic computer program. Since many of these test cases included *Journal* CPCs, cases for the trial were selected from 1969, a year from which no previous *Journal* cases had been presented to INTERNIST-I. Before entering any cases, project members serially reviewed the published final anatomic diagnoses. All cases in which one or more of the major diagnoses were not represented in INTERNIST-I's still incomplete knowledge base were rejected. The diagnostic program cannot conclude a diagnosis that is missing from the knowledge base; such a case would not be a fair test for the system. The excluded diagnoses were neither more rare nor more complex than the diagnoses chosen for analysis. Cases 1-1969 through 42-1969 (inclusive) were reviewed, and 19 cases were obtained in which all major CPC diagnoses were included in the data base. That only 19 of 42 cases reviewed qualified for the study is not unexpected. It is estimated that the current INTERNIST-I knowledge base includes roughly 70 to 75 per cent of the major diagnoses in internal medicine. If each case on the average contained three major diagnoses, the probability that all three diagnoses would be included in the knowledge base would be $(0.75) \times (0.75) \times (0.75)$ or 42 per cent.

In establishing criteria for evaluating performance on the *Journal* CPCs, one must classify final anatomic or clinical diagnoses as major or minor. Major diagnoses are defined as those central to the problem. Classified as minor diagnoses are diseases that were present in the patient but were clinically less relevant, including diseases only partially described in the published case protocol, or conditions that were successfully managed and that subsequently resolved. Diagnostic decisions made by the clinicians at the Massachusetts General Hospital, by the case discussants, and by INTERNIST-I were classified as correct when they were confirmed by the pathologists or when a clinical syndrome was universally agreed to be present. When either the physicians or INTERNIST-I introduced an incorrect diagnosis, a separate notation was made, because an incorrect diagnosis has a different meaning from that of a failure to make a correct diagnosis. We recognize two ways for a program or a clinician to make a correct diagnosis in the setting of a CPC: to state unequivocally that the patient has the disease ("definitive diagnosis") or to offer an unresolved differential diagnosis that includes the correct diagnosis as its topmost element ("tentative diagnosis"). INTERNIST-I makes definitive diagnoses by conclusion and tentative diagnoses by deferral (see above). The hospital clinicians and the case discussants also made both types of diagnoses. A tentative diagnosis was counted as incorrect if its topmost element was not the correct diagnosis, even if the associated differential diagnosis included the correct diagnosis.

Table 5 summarizes the results in the 19 trial cases. There were 43 possible correct major diagnoses. INTERNIST-I, the clinicians at the

Table 5. Summary of Results for Major Diagnoses in 19 Cases Used in the
INTERNIST-I Evaluation

Category	INTERNIST-I	Clinicians	Discussant
	No. of Instances		
Total possible diagnoses	*43*	*43*	*43*
Definitive, correct	17	23	29
Tentative, correct	8	5	6
Failed to make correct diagnosis	18	15	8
Definitive, incorrect	5	8	11
Tentative, incorrect	6	5	2
Total no. of incorrect diagnoses	11	13	13
Total no. of errors in diagnosis	29	28	21

Massachusetts General Hospital, and the case discussants made 17, 23,
and 29 correct definitive diagnoses, respectively. A correct tentative diagnosis was offered eight, five, and six times, respectively. Thus, of 43 anatomically verified diagnoses, INTERNIST-I failed to make a total of 18,
whereas the clinicians failed to make 15 such diagnoses and the discussants missed only eight. Of the 18 situations in which INTERNIST-I
failed to make an anatomically correct diagnosis, the clinicians or the discussant or both failed to make the correct diagnosis 11 times. INTERN-
IST-I made a correct diagnosis in seven circumstances in which the clinicians or the case discussant or both failed to do so. INTERNIST-I
made five incorrect definitive diagnoses and six incorrect tentative diagnoses (naming diseases that were not present in the patients). The Massachusetts General Hospital clinicians made eight incorrect definitive
diagnoses and five incorrect tentative diagnoses. The case discussants
made 11 incorrect definitive diagnoses and two incorrect tentative diagnoses. Four of the five situations in which INTERNIST-I made an incor-

Table 6. Classification of Errors Made by INTERNIST-I During the Evaluation

Type of Error	No. of Occurrences
Knowledge-base errors	
Data base incomplete/omission	2
Data base incorrect	2
Lack of anatomic knowledge	1
Failure to represent degree of severity	2
Computer-program faults	
Lack of temporal reasoning	3
Failure of scoring algorithm	3
Failure to seek global overview	1
Improper attribution of a finding to a concluded diagnosis	6

rect definitive diagnosis were situations in which the discussant also made a wrong diagnosis.

The shortcomings of the program, which were highlighted by the evaluation, fall into two general categories. The first type are limitations due to the structure or content of the knowledge base. Examples include the absence of a manifestation required to describe an important finding; the use of overly simplistic manifestations for some circumstances; the inadvertent omission of a finding from a disease profile; the assignment of an incorrect evoking strength, frequency, or import; and the failure of a manifestation to convey adequate anatomic information. The second type of limitation resulted from deficiencies in the design or implementation (or both) of the computer program. Included in this category were the failure to incorporate temporal reasoning capabilities; problems resulting from the use of the scoring algorithm; the inability to take a broad overview in attacking a complex problem; and the improper attribution of findings to concluded diagnoses (i.e., invoking the wrong explanation for a finding). Specific reasons for INTERNIST-I's incorrect diagnoses (made both by omission and by commission) are listed in Table 6.

Discussion

Experience with INTERNIST-I has reinforced our impression of medical diagnosis as a complex process. Diagnosis consists of two fundamental activities: the generation of one or more differential diagnoses (each for a separate problem area) and the resolution of individual differential diagnoses. The surprising ability of the program to make multiple and complex diagnoses in the broad field of internal medicine emphasizes the power of its underlying heuristic methods.

Several important shortcomings of the INTERNIST-I approach to diagnosis merit further investigation. Feinstein [12] has emphasized the importance of explanation as part of diagnostic reasoning. INTERNIST-I's greatest failing during the evaluation (occurring in six instances) was its inability to attribute findings to their proper causes. Because of the ad hoc, serial nature of INTERNIST-I's formation of problem areas, the program cannot synthesize a general overview in complicated multisystem problems. The structure of the knowledge base, especially the form of the disease profiles, limits the program's ability to reason anatomically or temporally. The program cannot recognize subcomponents of an illness, such as specific organ-system involvements or the degree of severity of pathologic processes.

A diagnostic program must be able to recognize the appropriate cause or causes of observed findings in a patient. A justification for each diagnosis must be developed on a pathophysiologic or causal framework that is consistent with established medical knowledge. To its detriment,

INTERNIST-I's handling of explanation is shallow. When the program concludes a diagnosis, that diagnosis is allowed to explain any observed manifestations that are listed on its disease profile. Once explained, a manifestation is no longer used to evoke new disease hypotheses or to participate in the scoring process. This situation is compounded by the inadequate representation of causality in the INTERNIST-I knowledge base. Disease profiles contain, in an undifferentiated manner, factors predisposing a patient to the illness as well as findings that result from the disease process itself. An example of this problem occurred in the analysis of Case 17-1969, when INTERNIST-I allowed hepatic encephalopathy to explain the finding of hypokalemia. The program should have recognized hypokalemia as a predisposing factor for hepatic encephalopathy and initiated a search for an independent cause of the finding. At present, the limitations of the knowledge base prohibit such activity.

What is required is a restructuring of the knowledge base to include intermediate-level pathophysiologic states and the segregation of predisposing factors from findings actually caused by a disease. Diseases should be profiled in terms of their intermediate states rather than as exhaustive lists of manifestations. If the program had such a feature, the presence or absence of each state would be independently determined, and a disease would be allowed to explain a finding only when the state causing the finding was confirmed.

A related problem not handled well by INTERNIST-I is the interdependency of manifestations. For example, persons with elevated conjugated bilirubin levels in their blood usually have bilirubinuria. At present, the evoking strengths of each finding count redundantly toward any diagnosis that can explain them. This phenomenon causes INTERNIST-I to favor disproportionately the most common explanation for a set of findings. A solution would be the creation of an intermediate-level state, "abnormal bilirubin metabolism and transport," which would explain both conjugated hyperbilirubinemia and bilirubinuria. Appropriate weight for the intermediate state (rather than for the interdependent manifestations) could be given to any diseases that cause it. Thus, the creation of a causal network of pathophysiologic states, interposed between observable manifestations and final diagnoses, would allow a diagnostic program to attribute findings to causes accurately and would help to diminish the influence of interdependent manifestations of disease.

INTERNIST-I constructs differential diagnoses in an ad hoc manner, using a scoring algorithm to define the topmost (best) diagnosis and another program, the partitioner, to define reasonable competitors for the topmost diagnosis. By formulating and focusing attention on only one problem domain at any given time, the program is able to disregard red herrings and to set aside temporarily findings caused by disease processes falling outside the selected problem domain. By creating and processing problem domains serially, the program is able to make multiple diagnoses. But INTERNIST-I cannot formulate a broad perspective in complicated multisystem problems. It is constrained to working with tunnel vision, discriminating among diagnoses within each problem area and remaining unable to look at several problem areas simultaneously. Only after a specific diagnosis is concluded can INTERNIST-I use the links in its data base to give bonus weight to related diagnoses in separate problem domains. New programming approaches to complex reasoning processes have been developed [8] to enable CADUCEUS, the successor to INTERNIST-I, to synthesize a broad overview incorporating causal relations into an approach to a patient's problems.

INTERNIST-I is unable to reason anatomically or temporally. The program could not differentiate gastric compression due to pancreatic mass effect from that due to hepatic mass effect in Case 23-1969, and in doing so it erroneously concluded that the patient had a hepatoma rather than pancreatitis. Nor can INTERNIST-I recognize the degree of severity of a finding or process in all instances. Two of INTERNIST-I's failures during the evaluation resulted from its inadequate recognition of the degree of severity of an individual manifestation (a decreased blood potassium level) and of an organ-system involvement by a pathologic process (disseminated vasculitis). Reorganization of the data base to allow representation of these concepts is also being undertaken.

INTERNIST-I is only one of many computer-based tools with the purpose of extending the capabilities of the physician. Such programs can broaden the clinician's scope and awareness of data for the diagnosis and treatment of illness. For the present, INTERNIST-I remains a research tool. After refinement of the knowledge base and diagnostic programs, a prospective clinical trial will be required to compare the program's behavior with that of clinicians in terms of diagnostic accuracy, cost effectiveness, and danger to the patient.

Appendix: A Sample Case Analysis

The transcript of an INTERNIST-I case analysis given below illustrates the operation of the diagnostic programs. The case was taken from a CPC published in the *Journal* in 1969 [11]. The laboratory values are reported as measured in 1969. The bracketed paragraphs labeled "Comment" have been interpolated for clarification; they are not part of the actual consultation. Places where the transcript has been abridged are indicated by an ellipsis.

INTERNIST-I consultation   SUMEX-AIM Version
15-May-81 07:31:39
ENTER CASE NAME: NEJM-CASE-30-1969-ADMISSION-1

[Comment: Here the user enters the initial positive findings (present in the patient) and negative findings (absent). The specialized INTERNIST-I vocabulary of some 3550 manifestations must be used in describing the case. The plus (+) prompt precedes each positive finding entered by the user. Because INTERNIST-I has no mechanism for the representation of time, all findings have been collapsed into a single list, independently of their order of appearance in the patient.]

INITIAL POSITIVE MANIFESTATIONS:
+AGE GTR THAN 55
+ARTHRITIS HX
+DEPRESSION HX
+SEX FEMALE
+THYROIDECTOMY HX
+ULCER PEPTIC HX
+URINE DARK HX
+WEIGHT INCREASE RECENT HX
+ANOREXIA
+CHEST PAIN LATERAL EXACERBATION WITH BREATHING
+CHEST PAIN LATERAL SHARP
+DYSPNEA ABRUPT ONSET
+ABDOMEN DISTENTION
+ABDOMEN FLUID WAVE
+ASTERIXIS
+FECES LIGHT COLORED
+JAUNDICE
+JOINT(S) PERIARTICULAR THICKENING
+JOINT(S) RANGE OF MOTION DECREASED
+LIVER ENLARGED MODERATE
+PLEURAL FRICTION RUB
+PULSE PRESSURE INCREASED
+SKIN PALMAR ERYTHEMA
+SKIN SPIDER ANGIOMATA
+SPLENOMEGALY SLIGHT
+TACHYCARDIA
+TACHYPNEA
+THYROID ENLARGED ASYMMETRICAL
+ALKALINE PHOSPHATASE BLOOD GTR THAN 2 TIMES NORMAL
+BILIRUBIN BLOOD CONJUGATED INCREASED
+BILIRUBIN URINE PRESENT
+FECES GUAIAC TEST POSITIVE
+GLUCOSE BLOOD 130 TO 300
+SGOT 120 TO 400
+AMMONIA BLOOD INCREASED
+ASCITIC FLUID PROTEIN 3 GRAM(S) PERCENT OR LESS
+ASCITIC FLUID WBC 100 TO 500
+ELECTROPHORESIS SERUM ALBUMIN DECREASED
+ELECTROPHORESIS SERUM BETA GLOBULIN INCREASED
+ELECTROPHORESIS SERUM GAMMA GLOBULIN INCREASED
+GLUCOSE TOLERANCE DECREASED
+IMMUNOELECTROPHORESIS SERUM IGA INCREASED
+IMMUNOELECTROPHORESIS SERUM IGM INCREASED
+IMMUNOELECTROPHORESIS SERUM IGG INCREASED
+LDH BLOOD INCREASED
+LE TEST POSITIVE
+LIVER RADIOISOTOPE SCAN IRREGULAR UPTAKE
+PROTHROMBIN TIME INCREASED
+RHEUMATOID FACTOR POSITIVE
+STOMACH BARIUM MEAL HIATAL HERNIA
+*DENY*
[Comment: The key word DENY indicates to INTERNIST-I that the user now
wants to enter negative findings. The minus (−) prompt precedes each negative
finding entered.]

INITIAL NEGATIVE FINDINGS:
- −ALCOHOLISM CHRONIC HX
- −DIARRHEA CHRONIC
- −FECES BLACK TARRY
- −FEVER
- −PRESSURE VENOUS INCREASED ON INSPECTION
- −HEMATOCRIT BLOOD LESS THAN 35
- −UREA NITROGEN BLOOD 30 TO 59
- −URIC ACID BLOOD INCREASED
- −ASCITIC FLUID AMYLASE INCREASED
- −ASCITIC FLUID CYTOLOGY POSITIVE
- −ASCITIC FLUID LDH GTR THAN 500
- −ESOPHAGUS BARIUM MEAL VARICES
- −STOMACH BARIUM MEAL ULCER CRATER
- −T3 RESIN UPTAKE INCREASED
- −T4 TOTAL BLOOD INCREASED
- −*GO*

[Comment: The user enters *GO* to indicate that all relevant positive and negative findings have been entered. The INTERNIST-I consultant programs guide the user through the rest of the diagnostic workup by asking questions. For clarity, all responses typed by the user from this point on will be in italics.]
DISREGARDING: DEPRESSION HX, WEIGHT INCREASE RECENT HX, CHEST PAIN LATERAL EXACERBATION WITH BREATHING, CHEST PAIN LATERAL SHARP, DYSPNEA ABRUPT ONSET, ABDOMEN FLUID WAVE, ASTERIXIS, JOINT(S) PERIARTICULAR THICKENING, PLEURAL FRICTION RUB, THYROID ENLARGED ASYMMETRICAL, FECES GUAIAC TEST POSITIVE, GLUCOSE BLOOD 130 TO 300, AMMONIA BLOOD INCREASED, ASCITIC FLUID PROTEIN 3 GRAM(S) PERCENT OR LESS, ASCITIC FLUID WBC 100 TO 500, GLUCOSE TOLERANCE DECREASED
CONSIDERING: AGE GTR THAN 55, SEX FEMALE, URINE DARK HX, ANOREXIA, FECES LIGHT COLORED, JAUNDICE, LIVER ENLARGED MODERATE, SKIN PALMAR ERYTHEMA, SKIN SPIDER ANGIOMATA, SPLENOMEGALY SLIGHT, ALKALINE PHOSPHATASE BLOOD GTR THAN 2 TIMES NORMAL, BILIRUBIN BLOOD CONJUGATED INCREASED, BILIRUBIN URINE PRESENT, SGOT 120 TO 400, ELECTRO-PHORESIS SERUM ALBUMIN DECREASED, ELECTROPHORESIS SERUM GAMMA GLOBULIN INCREASED, IMMUNOELECTROPHORESIS SERUM IGA INCREASED, IMMUNOELECTROPHORESIS SERUM IGG INCREASED, IMMUNOELECTROPHORESIS SERUM IGM INCREASED, LDH BLOOD INCREASED, LE TEST POSITIVE, LIVER RADIOISOTOPE SCAN IRREGULAR UPTAKE, PROTHROMBIN TIME INCREASED, RHEUMATOID FACTOR POSITIVE DISCRIMINATE: HEP-ATITIS CHRONIC ACTIVE, BILIARY CIRRHOSIS PRIMARY

[Comment: At this point, INTERNIST-I has constructed a master differential-diagnosis list, ordered its members with a scoring algorithm, and then focused attention on the most promising problem area, which appropriately contains two liver diseases. The DISREGARDING list consists of all findings that are inconsistent with the topmost (first-listed) diagnosis; the CONSIDERING list includes all findings explained by the topmost diagnosis (i.e., chronic active hepatitis). The number of plausible contenders in the problem area determines the strategy for questioning. Questions are asked to discriminate between the two diagnoses.]
Please Enter Findings of LIPID(S) BLOOD
GO

CHOLESTEROL BLOOD DECREASED?
N/A
[Comment: INTERNIST-I asks questions in two forms. A question of the style "Please Enter Findings of ..." asks about a generic class of findings, such as serum lipids or chest x-ray films. The user enters specific (positive or negative) findings and when finished types *"GO."* If there are no positive or negative findings to enter, typing *"GO"* will cause the programs to ask the original single item that prompted the generic question. For example, an elevated cholesterol level would favor primary biliary cirrhosis over chronic active hepatitis, whereas the converse situation would apply if the cholesterol were decreased. Thus, INTERNIST-I asked a generic question about serum lipids. The second form of questioning is direct. INTERNIST-I simply prints a manifestation name followed by a question mark. The user can then respond by typing *"YES"* (present), *"NO"* (absent), or *"N/A"* (not available—either unknown or not performed). The published CPC did not mention the serum cholesterol value, so the user entered *N/A.*]
Please Enter Findings of INSPECTION AND PALPATION SKIN
GO
SKIN XANTHOMATA?
NO
Please Enter Findings of BLOOD COUNT AND SMEAR
GO
RBC TARGET(S)
NO
   [Comment: INTERNIST-I uses information about the type of each manifestation—history, symptoms, signs, and laboratory tests, both routine and two gradations of more expensive or invasive laboratory procedures. The types of manifestations determine which questions are asked. Although the greatest discriminating power for resolving a differential diagnosis often resides in an invasive test, such as a liver biopsy, INTERNIST-I has been programmed to exhaust the useful history and physical-examination findings before escalating its questioning to more expensive or dangerous laboratory tests.]
Please Enter Findings of KIDNEY FUNCTION TEST(S)
GO
UREA NITROGEN BLOOD 60 TO 100?
NO
UREA NITROGEN BLOOD GTR THAN 100?
NO
Please Enter Findings of LIVER FUNCTION TEST(S)
GO
UROBILINOGEN URINE ABSENT?
N/A
   [Comment: INTERNIST-I did not seek to investigate renal function itself in the passage above. It was programmed to recognize that the urinary urobilinogen value may be unreliable in the presence of marked azotemia, so it first asked about the blood urea nitrogen.]
CONCLUDE: HEPATITIS CHRONIC ACTIVE
   [Comment: It is likely that INTERNIST-I had nearly made the diagnosis of chronic active hepatitis before it asked any questions. The negative answers favored chronic active hepatitis, and therefore INTERNIST-I concluded the diagnosis. The programs now recycle, using the findings not explained by the concluded diagnosis.]
DISREGARDING: DEPRESSION HX, CHEST PAIN LATERAL EXACERBATION WITH BREATHING, CHEST PAIN LATERAL SHARP, DYSPNEA

ABRUPT ONSET, ASTERIXIS, JOINT(S) PERIARTICULAR THICKENING, PLEURAL FRICTION RUB, THYROID ENLARGED ASYMMETRICAL, FECES GUAIAC TEST POSITIVE, GLUCOSE BLOOD 130 TO 300, AMMONIA BLOOD INCREASED, GLUCOSE TOLERANCE DECREASED
CONSIDERING: WEIGHT INCREASE RECENT HX, ABDOMEN DISTENTION, ABDOMEN FLUID WAVE, ASCITIC FLUID OBTAINED BY PARACENTESIS, ASCITIC FLUID WBC 100 TO 500, ASCITIC FLUID PROTEIN 3 GRAM(S) PERCENT OR LESS
. . .
CONCLUDE: TRANSUDATIVE ASCITES
DISREGARDING: . . .
CONSIDERING: ASTERIXIS, PULSE PRESSURE INCREASED, TACHY-
    CARDIA, AMMONIA BLOOD INCREASED
PURSUING: HEPATIC ENCEPHALOPATHY
    [Comment: The links in the INTERNIST-I data base between chronic active hepatitis and hepatic encephalopathy have resulted in a bonus weight's being given to hepatic encephalopathy here; previously, links had promoted the consideration of transudative ascites, since it can also be caused by chronic active hepatitis.]
CSF FLUID OBTAINED?
N/A
    [Comment: Here INTERNIST-I was about to ask about the glutamine level in the cerebrospinal fluid. Since no lumbar puncture was performed, the result is not available.]
CONCLUDE: HEPATIC ENCEPHALOPATHY
    [Comment: In the above situation, there were no diagnostically helpful tests remaining for INTERNIST-I to ask. INTERNIST-I has been programmed to relax its criteria for concluding a diagnosis when all useful lines of questioning have been blocked. Since INTERNIST-I had been close to making the diagnosis of hepatic encephalopathy, the program now concludes the diagnosis. The case analysis was intentionally stopped at this point, because all relevant major diagnoses had been covered. Without such intervention, INTERNIST-I would try to explain any remaining important findings, such as the arthritis and pleurisy.]

Acknowledgment. The INTERNIST-I/CADUCEUS project is supported by grants from the Division of Research Resources (R24 RR 01101) and the National Library of Medicine (R01 LM 03710 and R23 LM03589), National Institutes of Health. The SUMEX Computing Project is supported by a grant (RR 00785) from the Biotechnology Resources Program, National Institutes of Health.

We are indebted to Craig D. Dean, Charles E. Oleson, and Kenneth W. Quayle for their contributions in writing the INTERNIST-I computer programs; to Zachary C. Moraitis for his assistance in the conceptual design of the project and in the development of the knowledge base; to a large number of medical students and several fellows in computer medicine for their assistance in the development of the INTERNIST-I knowledge base; and to the staff of the SUMEX-AIM computing facility of the National Institutes of Health for providing expert assistance and a friendly environment for programming.

References

[1] Wardle A, Wardle L. Computer aided diagnosis—a review of research. *Methods Inf Med.* 1978; 17:15–28.

[2] Wagner G, Tautu P, Wolber U. Problems of medical diagnosis—a bibliography. *Methods Inf Med.* 1978; 17:55–74.

[3] Shortliffe EH, Buchanan BG, Feigenbaum EA. Knowledge engineering for medical decision making: a review of computer-based clinical decision aids. *Proc IEEE.* 1979; 67:1207–24.

[4] de Dombal FT, Leaper DJ, Staniland JR, McCann AP, Horrocks JC. Computer-aided diagnosis of abdominal pain. *Br Med J.* 1972; 2:9–13.

[5] Yu VL, Fagan LM, Wraith SM, et al. Antimicrobial selection by computer. JAMA. 1979; 242:1279–82.

[6] Pople HE, Myers JD, Miller RA. DIALOG: a model of diagnostic logic for internal medicine. In: Proceedings of the Fourth International Joint Conference on Artificial Intelligence. Cambridge, Mass.: MIT Artificial Intelligence Laboratory Publications, 1975:848–55.

[7] Myers JD, Pople HE, Miller RA. INTERNIST: can artificial intelligence help? In: Connelly, Benson, Burke, Fenderson, eds. Clinical decisions and laboratory use. Minneapolis: University of Minnesota Press, 1982:251–69.

[8] Pople HE. Heuristic methods for imposing structure on ill-structured problems: the structuring of medical diagnostics. In: Szolovits P, ed. Artificial intelligence in medicine, AAAS Symposium Series, Boulder, Colo.: Westview Press, 1982:119–85.

[9] Winston PH. Artificial intelligence. Reading, Mass.: Addison-Wesley, 1977.

[10] Nilsson NJ. Principles of artificial intelligence. Palo Alto: Tioga Publishing Co., 1980.

[11] Case Records of the Massachusetts General Hospital (Case 30-1969). *N Engl J Med.* 1969; 281:206–13.

[12] Feinstein AR. Clinical biostatistics XXXIX: the haze of Bayes, the aerial palaces of decision analysis, and the computerized Ouija board. *Clin Pharmacol Ther.* 1977; 21:482–96.

23

Diagnostic Expert Systems Based on a Set Covering Model

James A. Reggia, Dana S. Nau, and Pearl Y. Wang

This paper proposes that a generalization of the set covering problem can be used as an intuitively plausible model for diagnostic problem solving. Such a model is potentially useful as a basis for expert systems in that it provides a solution to the difficult problem of multiple simultaneous disorders. We briefly introduce the theoretical model and then illustrate its application in diagnostic expert systems. Several challenging issues arise in adopting the set covering model to real-world problems, and these are also discussed along with the solutions we have adopted.

1. Introduction

A diagnostic problem can be defined to be a problem in which one is given a set of abnormal findings (manifestations) for some system, and must explain why those findings are present. Problems of this kind are very common: they include diagnosing a patient's signs and symptoms, determining why a computer program failed, deciding why an automobile will not start, finding the cause of noise in a plumbing system, localizing a fault in an electronic circuit, etc. Because of this ubiquity, developing general methods for expert systems which support the decision making of human diagnosticians is an important issue at present.

This paper introduces a new model for diagnostic expert systems based on the concept of minimal set covers. This model is of interest because it captures several intuitively plausible features of human diagnostic inference, it directly addresses the problem of multiple simultaneous causative disorders, and it provides a basis for a theory of diagnostic inference.

Reprinted with permission from the *International Journal of Man-Machine Studies, 19*, pp. 437–460. Copyright 1983 by Academic Press Inc. (London) Limited.

In the following, section 2 discusses the set covering model, and section 3 explains how the model can be adopted for use in expert diagnostic systems. Section 4 and Appendix B give examples of operational expert systems based on set covering, and section 5 presents in a more detailed fashion some of the issues involved in implementing these systems. Section 6 contains some concluding remarks.

2. The Set Covering Model

In the set covering model the underlying knowledge for a diagnostic problem is organized as pictured in Figure 1(a) (a table of symbols is given in Appendix A). There are two discrete finite sets which define the scope of diagnostic problems: D, representing all possible *disorders* d_i that can occur, and M, representing all possible *manifestations* m_j that may occur when one or more disorders are present. For example, in medicine, D might represent all known diseases (or some relevant subset of all diseases, see below), and M would then represent all possible symptoms, examination findings, and abnormal laboratory results that can be caused by diseases in D. We will assume that $D \cap M = \emptyset$.

To capture the intuitive notion of causation, we assume knowledge of a relation $C \subseteq D \times M$, where $\langle d_i, m_j \rangle \in C$ represents "d_i can cause m_j." Note that $\langle d_i, m_j \rangle \in C$ does not imply that m_j always occurs when d_i is present, but only that m_j may occur. For example, a patient with a heart attack may have chest pain, numbness in the left arm, loss of consciousness, or any of several other symptoms, but none of these symptoms are necessarily present.

Given D, M, and C, the following sets can be defined:

$$man(d_i) = \{m_j \mid \langle d_i, m_j \rangle \in C\} \quad \forall d_i \in D, \text{ and}$$
$$causes(m_j) = \{d_i \mid \langle d_i, m_j \rangle \in C\} \quad \forall m_j \in M.$$

These sets are depicted in Figure 1(a), and represent all possible manifestations caused by d_i, and all possible disorders that cause m_j, respectively. These concepts are intuitively familiar to the human diagnostician. For example, medical textbooks frequently have descriptions of diseases

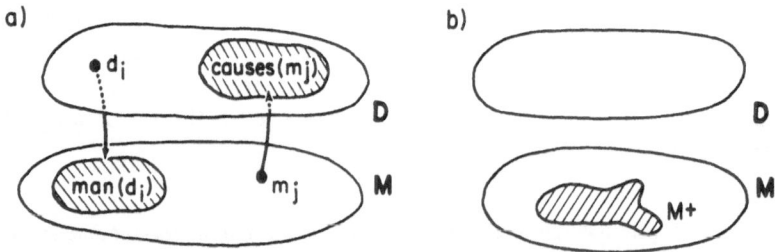

Figure 1. Organization of diagnostic knowledge (a) and problems (b).

which include, among other facts, the set man(d_i) for each disease d_i. Physicians often refer to the "differential diagnosis" of a symptom, which corresponds to the set causes(m_j). Clearly, if man(d_i) is known for every disorder d_i, or if causes(m_j) is known for every manifestation m_j, then the causal relation C is completely determined. We will use man(D) = $\cup_{d_i \in D}$ man(d_i) and causes (M) = $\cup_{m_j \in M}$ causes(m_j) to indicate all possible manifestations of a set of disorders D and all possible causes of any manifestation in M, respectively.

Finally, there is a distinguished set $M^+ \subseteq M$ which represents those manifestations which are known to be present (see Figure 1(b)). Whereas **D**, **M**, and **C** are general knowledge about a class of diagnostic problems, M^+ represents the manifestations occurring in a specific case.

Using this terminology, we can now make the following definition.

Definition. A *diagnostic problem P* is a 4-tuple $\langle$**D**, **M**, **C**, $M^+ \rangle$ where these components are as described above.

We will assume in what follows that diagnostic problems are well-formed in the sense that man(d_i) and causes(m_j) are always non-empty sets.

Having characterized a diagnostic problem in these terms, we now turn to defining the solution to a diagnostic problem by first introducing the concept of explanation.

Definition. For any diagnostic problem P, $E \subseteq$ **D** is an *explanation* for M^+ if; (i) $M^+ \subseteq$ man(E), or in words: E *covers* M^+; and (ii) $|E| \leq |D|$ for any other cover D of M^+, i.e., E is *minimal*.

This definition captures what one intuitively means by "explaining" the presence of a set of manifestations. Part (i) specifies the reasonable constraint that a set of disorders E must be able to cause all known manifestations M^+ in order to be considered an explanation for those manifestations. However, that is not enough: part (ii) specifies that E must also be one of the smallest sets to do so. Part (ii) reflects the Principle of Parsimony or Ockham's Razor: the simplest explanation is the preferable one. This principle is generally accepted as valid by human diagnosticians. Here, we have equated "simplicity" with minimal cardinality, reflecting an underlying assumption that the occurrence of one disorder d_i is independent of the occurrence of another.

With these concepts in mind, we can now define the solution to a diagnostic problem.

Definition. The *solution* to a diagnostic problem P, designated Sol(P), is the set of all explanations for M^+.

The concepts defined above are illustrated in the following example.

Example. Let P = $\langle$**D**, **M**, **C**, $M^+ \rangle$ where **D** = $\{d_1, d_2, \ldots, d_9\}$, **M** = $\{m_1, \ldots, m_6\}$, and man(d_i) and causes(m_j) are as specified in Table 1. Note

Table 1. Knowledge About a Class of
Diagnostic Problems

d_i	$man(d_i)$
d_1	$m_1\ m_4$
d_2	$m_1\ m_3\ m_4$
d_3	$m_1\ m_3$
d_4	$m_1\ m_6$
d_5	$m_2\ m_3\ m_4$
d_6	$m_2\ m_3$
d_7	$m_2\ m_5$
d_8	$m_4\ m_5\ m_6$
d_9	$m_2\ m_5$

m_j	$causes(m_j)$
m_1	$d_1\ d_2\ d_3\ d_4$
m_2	$d_5\ d_6\ d_7\ d_9$
m_3	$d_2\ d_3\ d_5\ d_6$
m_4	$d_1\ d_2\ d_5\ d_8$
m_5	$d_7\ d_8\ d_9$
m_6	$d_4\ d_8$

The relation C is implicitly defined by either
the top or bottom half of this table

that the top (or bottom) half of Table 1 implicitly defines the relation C, because $C = \{\langle d_i, m_j\rangle\,|\,m_j \in man(d_i)$ for some $d_i\}$. Let $M^+ = \{m_1, m_4, m_5\}$. Note that no single disorder can cover (account for) all of M^+, but that some pairs of disorders do cover M^+. For instance, if $D = \{d_1, d_7\}$ then $M^+ \subseteq man(D)$. Since there are no covers for M^+ of smaller cardinality than D, it follows that D is an explanation for M^+. Careful examination of Table 1 should convince the reader that

$$Sol(P) = \{\{d_1\ d_7\}\{d_1\ d_8\}\{d_1\ d_9\}\{d_2\ d_7\}\{d_2\ d_8\}\{d_2\ d_9\}\{d_3\ d_8\}\{d_4\ d_8\}\}$$

is the set of all explanations for M^+.

It is of interest to compare the model of diagnostic problems presented here with the classic set covering problem. The set covering problem is typically stated along the following lines (Edwards, 1962):

> For a finite set S of elements and a family F of subsets of S, a cover K of S from F is a subfamily $K \subseteq F$ such that $\cup(K) = S$. A cover K is called minimum if its cardinality is as small as possible.

In this definition, S corresponds to M^+ and F corresponds to **D** in the sense that each $d_i \in \mathbf{D}$ labels a subset of M^+ (the intersection of $man(d_i)$ with M^+). A minimum cover K corresponds roughly to the idea of an explanation E except $man(E)$ is required only to contain M^+ rather than be equal to M^+.

3. Expert Systems Using the Set Covering Model

We now turn to the description of expert systems for diagnostic problem solving based on the set covering model presented above. Such systems are organized as shown in Figure 2 and consist of three parts.

1. A *database,* which is divided into case-specific information and general knowledge about some domain of diagnostic problems. We will use the term *knowledge base* for the latter.
2. An *inference mechanism* which is a hypothesize-and-test process that mimics diagnostic reasoning by using the set covering model.
3. A *user interface* which accepts assertions and queries from the user and translates them into internal data structures.

We now present a specific example of an expert system called System D for diagnostic problem solving. This implemented system illustrates how the set cover model can be adopted to the demands of real world problems. While System D is medically oriented, it should be remembered that the set cover model is domain-independent and not restricted to problems of medical diagnosis.

System D is a relatively large expert system for diagnosing patients with dizziness. Dizziness is in general a very difficult diagnostic problem for the physician because there are numerous potential causes that are distributed across multiple medical specialties. Examples of possible diagnoses include:

orthostatic hypotension secondary to drugs (orthostatic hypotension is a fall in blood pressure upon standing up, and can be a side effect of certain medications);

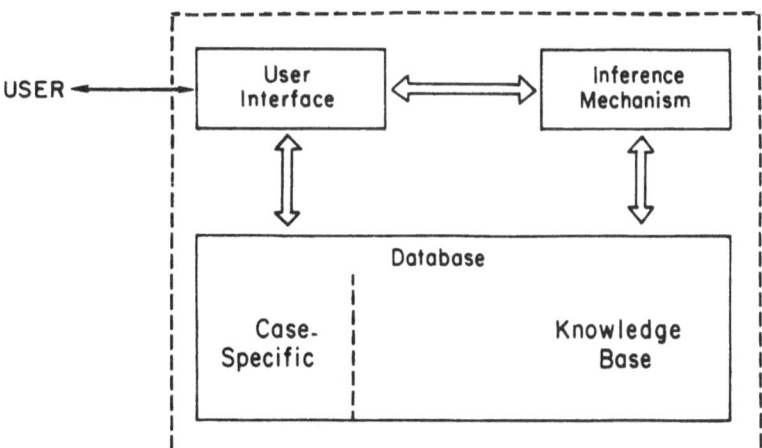

Figure 2. Architecture of an expert system based on the set covering model.

heart disease, such as an irregular heart beat or an abnormal heart valve;
basilar migraine: headache due to painfully dilated blood vessels which supply blood to the balance centers of the brain;
inner ear diseases: these interfere with the balance mechanisms of the inner
ear, and include viral labyrinthitis, Meniere's disease, and otosclerosis; and
hyperventilation: overbreathing, typically secondary to anxiety.

It is entirely possible that more than one cause of dizziness could be present simultaneously.

The knowledge base for System D is derived from numerous references and currently contains information about 50 causes of dizziness. It
was built using KMS, a domain-independent software facility for constructing expert systems (Reggia, 1981). We now describe its components
in detail.

The Case-Specific Database

The case-specific database for System D contains a collection of assertions that describe a specific diagnostic problem. For example,

AGE = 50;

DIZZINESS = PRESENT
  [TYPE = VERTIGO; COURSE = EPISODIC]; and

NEUROLOGICAL SYMPTOMS = DIPLOPIA

represent three assertions that might appear in the database. Each assertion is of the form

attribute relation value [elaboration],

so the three statements here mean: "A 50-year-old individual with episodic vertigo (a type of dizziness where one feels a sensation of motion)
and double vision (diplopia)." During a problem solving session this
case-specific information is acquired in a sequential fashion, generally in
response to questions generated by the expert system. The legal attributes
and their possible values are predefined in a database schema by the creator of the expert system (the "knowledge base author").

The Knowledge Base: Representing Diagnostic Knowledge

One of the attractive features of the set cover model is that it permits the
organization of diagnostic knowledge in a form familiar to the human
diagnostician. Information in the knowledge base is organized into
frame-like entities called DESCRIPTIONs. Each DESCRIPTION provides a textbook-like summary of the disorder with which it is associated.
An example of a DESCRIPTION from System D is illustrated in Figure
3. To understand this descriptive knowledge representation more fully, it

```
MENIERE'S DISEASE ⟨L⟩
  [DESCRIPTION:
    AGE = FROM 20 TO 30 ⟨L⟩;
    DIZZINESS = PRESENT
      [TYPE = VERTIGO;
       COURSE = ACUTE AND PERSISTENT,
       EPISODIC [EPISODE DURATION = MINUTES ⟨L⟩, HOURS ⟨H⟩;
                 OCCURRENCE = POSITIONAL ⟨H⟩, ORTHOSTATIC ⟨M⟩,
                          NON-SPECIFIC ⟨L⟩]    ];
    HEAD PAIN = PRESENT ⟨L⟩ [PREDOMINANT LOCATION = PERIAURAL];
    NEUROLOGIC SYMPTOMS = HEARING LOSS BY HISTORY ⟨H⟩, TINNITUS ⟨H⟩;
    PULSE DURING DIZZINESS = MARKED TACHYCARDIA ⟨L⟩;
    NEUROLOGIC SIGNS = NYSTAGMUS [TYPE = HORIZONTAL, ROTATORY],
                       IMPAIRED HEARING ⟨H⟩                        ]
```

Figure 3. The knowledge base of System D currently consists of a set of data structures called DESCRIPTIONs such as that shown here for Meniere's Disease.

is necessary to know about three conventions being used: symbolic probabilities, separation of causal and non-causal associations, and elaboration.

Symbolic probabilities, indicated in angular brackets in Figure 3, are subjective, non-numeric estimates of how frequently an event occurs. While exact probabilities of diagnostic associations are usually not available, a great deal of descriptive information about diagnosis exists in the form

> x *frequently* causes y,
> x *can* cause y,
> x is *never* associated with y,
> x is *commonly* associated with y,
> x is *rare* (*common, very common,* . . .), and
> x *only* occurs if y,

where x is some disorder and y is some fact about a case. Symbolic probabilities capture this coarse but useful information. The five† possible estimates we use are:

$$A = \underline{a}lways,$$
$$H = \underline{h}igh \ likelihood,$$
$$M = \underline{m}edium \ likelihood,$$
$$L = \underline{l}ow \ likelihood, \ and$$
$$N = \underline{n}ever.$$

---

†Five is obviously somewhat arbitrary, but it has proven sufficient for our applications so far.

Thus, the "⟨L⟩" following MENIERE'S DISEASE in Figure 3 indicates that this disorder is relatively uncommon, and the "⟨H⟩" on the last line of the DESCRIPTION indicates that Meniere's disease often causes impaired hearing.

The second convention used in DESCRIPTIONs is the separation of causal and non-causal associations. Certain features of a disorder can be viewed as being caused by the disorder being described. For example, in medicine loss of vision, chest pain, dizziness and confusion are all abnormalities that conceptually are caused by some underlying problem. We have been using the term *manifestations* for these causally-related features. In contrast, other features of a disorder are not causally associated with it. For example, a patient's age and sex may provide very significant information about the likelihood of a certain disease being present, but they are not caused by that disease. Features such as these will be referred to as *setting factors*. Which features in a knowledge base are manifestations and which are setting factors are indicated in the database schema specified by the knowledge base author (see Reggia, 1981). In the DESCRIPTION in Figure 3, only the first assertion concerning age specifies a setting factor, while each of the other assertions specify manifestations.

Finally, elaboration provides further details about a manifestation and is indicated as part of an assertion inside of square brackets. For example, in Figure 3 the assertion

```
DIZZINESS = PRESENT
  [TYPE = VERTIGO;
  COURSE = ACUTE AND PERSISTENT,
          EPISODIC . . . ]
```

elaborates on the type of dizziness manifested by Meniere's Disease by indicating that it is vertiginous in nature and that it occurs either in an acute, persistent fashion or in episodes.

With the above conventions in mind, the DESCRIPTION in Figure 3 should now be relatively understandable. It indicates that Meniere's disease is a relatively uncommon cause of dizziness (because of "⟨L⟩" immediately following the name of the disease). The dizziness it causes is vertiginous in nature and either acute and persistent or episodic. When episodic, the episodes usually last for hours and are especially produced by positional changes of the head. Meniere's Disease occasionally causes periaural headache, frequently causes hearing loss and tinnitus (ringing in the ears), and so forth.

What is most important here in the context of the set covering model is that the DESCRIPTION associated with any disorder d_i specifies, among other things, the set man(d_i) of all manifestations caused by d_i. Thus, a knowledge base containing a set of disorders along with all of their DESCRIPTIONs completely specifies the information needed to solve diagnostic problems as they were defined earlier. Returning to our example, the knowledge base of System D consists of a listing of 50 causes of dizziness and their DESCRIPTIONs, each similar to that illustrated in Figure 3. The knowledge base thus explicitly specifies the set **D** of all causative disorders of dizziness as well as the set man(d_i) for each disorder. Furthermore, the set **M** is implicitly specified, as it consists of every manifestation listed in any of the DESCRIPTIONs. The relationship **C** and the sets causes(m_j) are also implicitly specified by the collective information in the DESCRIPTIONs. As explained earlier, the DESCRIPTIONs in this knowledge base contain additional information about setting factors and estimates of relevant probabilities.

The Inference Mechanism: A Sequential Hypothesize-and-Test Process

In adapting the set covering model for use in a real-world expert system several issues must be addressed and resolved. Perhaps the most obvious of these issues is the fact that diagnostic problem-solving is inherently sequential in nature. Rather than knowing all of the manifestations which are present in a specific case at the start, the human diagnostician usually begins knowing that one or a few manifestations are present, and must actively seek further information about others. In medicine, for example, the physician is typically confronted with a patient complaining of some symptom (the "chief complaint"), and must uncover other manifestations through questions, examining the patient, and laboratory-testing.

Empirical studies done over the last decade have provided convincing evidence that this sequential diagnostic reasoning is guided by a hypothesize-and-test process (Elstein, Schulman & Sprafka, 1978; Kassirer & Gorry, 1978; see Reggia, 1982, for a review). Given a few initial manifestations, the human diagnostician constructs a tentative hypothesis about the cause of those manifestations. Further information is then sought for generally two reasons: either for completeness (so-called "protocol-driven" questions), or to uncover facts specifically needed to modify the evolving hypothesis (so-called "hypothesis-driven" questions). These latter questions "test" the validity of the hypothesis, possibly confirming or eliminating part of it.

This sequential diagnostic process can be captured in terms of the set covering model presented earlier. The tentative hypothesis at any point during problem-solving is defined to be the solution for those manifestations already known to be present, assuming, perhaps falsely, that no additional manifestations will be subsequently discovered. To construct and maintain a tentative hypothesis like this, three simple data structures prove useful:

MANIFS: the set of manifestations know to be present so far;
SCOPE: causes(MANIFS), the set of all disorders d_i for which at least one manifestation is already known to be present; and

FOCUS: the tentative solution for just those manifestations already in
          MANIFS; FOCUS is presented as a collection of generators.

The term "generator" used here needs further definition. Rather than representing the solution to a diagnostic problem as an explicit list of all possible explanations for M^+ or MANIFS, it is advantageous to represent the disorders involved as a collection of explanation generators. An explanation *generator* is a collection of sets of "competing" disorders that implicitly represent a set of explanations in the solution and can be used to generate them. A generator is analogous to a Cartesian set product, the difference being that the generator produces unordered sets rather than ordered tuples. To illustrate this idea, consider the example diagnostic problem presented earlier (Table 1). Two generators are sufficient to represent the solution to that problem: $\{d_1\ d_2\} \times \{d_7\ d_8\ d_9\}$ and $\{d_3\ d_4\} \times \{d_8\}$. The second generator here implicitly represents the two explanations $\{d_3\ d_8\}$ and $\{d_4\ d_8\}$, while the first generator represents the other six explanations in the solution.

There are at least three advantages to representing the solution to a diagnostic problem as a set of generators. First, this is usually a more compact form of the explanations present in the solution. Second, generators are a very convenient representation for developing algorithms to process explanations sequentially (see below). Finally, and perhaps most important, generators are closer to the way the human diagnostician organizes the possibilities during problem solving (i.e., the "differential diagnosis").

Using the three data structures MANIFS, SCOPE and FOCUS, a hypothesize-and-test algorithm based on the set covering model can perform diagnostic problem solving. The FOCUS represents the tentative or working hypothesis at any point during problem-solving. The algorithm, described informally, is:

1. Get the next manifestation m_j.
2. Retrieve causes(m_j) from the knowledge base.
3. MANIFS ← MANIFS ∪ $\{m_j\}$.
4. SCOPE ← SCOPE ∪ causes(m_j).
5. Adjust FOCUS to accommodate m_j.
6. Repeat this process until no further manifestations remain.

Thus, as each manifestation m_j that is present is discovered, MANIFS is updated simply by adding m_j to it. SCOPE is augmented to include any possible causes d_i of m_j which are not already contained in it (derived by taking the union of causes(m_j) and SCOPE). Finally, FOCUS is adjusted to accommodate m_j based on intersecting causes(m_j) with the sets of diseases in the existing generators. These latter operations are done such that any explanation which can no longer account for the augmented MANIFS (which now includes m_j) are eliminated.

The key step in this process is Step 5, the adjustment of the FOCUS or working hypothesis. Perhaps the best way to understand this step is to follow a simple example. Recall the abstract knowledge base illustrated in Table 1, and consider the same diagnostic problem $M^+ = \{m_1\ m_4\ m_5\}$ that was used earlier. The order in which information about manifestations is discovered is determined by question generation heuristics, as described later in section 5. For now, suppose that the sequence of events occurring during problem-solving were ordered as listed in Figure 4. What happens during problem-solving is as follows.

Initially, MANIFS, SCOPE and FOCUS are all empty. When m_1 is discovered to be present, m_1 is added to MANIFS, and the new SCOPE is the union of the old SCOPE with causes(m_1). Since previously there were no generators in the FOCUS, the intersection of causes(m_1) with them is trivially empty. In such situations a new generator is created, in this case consisting of causes(m_1). In the terms defined earlier, this generator represents a solution for $M^+ = \{m_1\}$. It tentatively postulates that there are four possible explanations for M^+, any one of which consists of a single disease. The FOCUS thus asserts that "d_1 or d_2 or d_3 or d_4 is present."

The absence of m_2 and m_3 do not change this initial hypothesis. However, when m_4 is discovered to be present, MANIFS and SCOPE are augmented appropriately. A new FOCUS is developed, representing the intersection of causes(m_4) with the single set in the only pre-existing generator set in FOCUS. Note that the new generator $\{d_1\ d_2\}$ in the FOCUS that results from this intersection operation represents precisely all explanations for the augmented MANIFS. This new FOCUS also illustrates another important point. As information about each possible manifestation becomes available, the FOCUS changes incrementally with a mono-

Events in order of their discovery	MANIFS	SCOPE	FOCUS
Initially	$\varnothing$	$\varnothing$	$\varnothing$
m_1 present	$\{m_1\}$	$\{d_1\ d_2\ d_3\ d_4\}$	$\{d_1\ d_2\ d_3\ d_4\}$
m_2 absent	$\{m_1\}$	$\{d_1\ d_2\ d_3\ d_4\}$	$\{d_1\ d_2\ d_3\ d_4\}$
m_3 absent	$\{m_1\}$	$\{d_1\ d_2\ d_3\ d_4\}$	$\{d_1\ d_2\ d_3\ d_4\}$
m_4 present	$\{m_1\ m_4\}$	$\{d_1\ d_2\ d_3\ d_4\ d_5\ d_8\}$	$\{d_1\ d_2\}$
m_5 present	$\{m_1\ m_4\ m_5\}$	$\{d_1\ d_2\ d_3\ d_4\ d_5\ d_7\ d_8\ d_9\}$	$\{d_1\ d_2\} \times \{d_7\ d_8\ d_9\}$ and $\{d_8\} \times \{d_3\ d_4\}$
m_6 absent	$\{m_1\ m_4\ m_5\}$	$\{d_1\ d_2\ d_3\ d_4\ d_5\ d_7\ d_8\ d_9\}$	$\{d_1\ d_2\} \times \{d_7\ d_8\ d_9\}$ and $\{d_8\} \times \{d_3\ d_4\}$

Figure 4. Sequential problem-solving using the set covering model.

tonic decrease in the number of explanations it represents (with the exception of situations where the FOCUS becomes empty).

When m_5 is noted to be present, MANIFS and SCOPE are again adjusted appropriately. However, in this case the intersection of causes(m_5) with the single generator set in the FOCUS is empty (none of the previous explanations represented by the old FOCUS can now cover all known manifestations). The occurrence of an empty FOCUS like this again triggers a restructuring of the FOCUS: a procedure is called that produces a new set of generators from the now augmented MANIFS and SCOPE. These new generators are based on the fact that the cardinality of any explanation now contained in the FOCUS must be exactly one greater than the cardinality of its old explanations. Thus, when m_5 is found to be present, the new generators represent explanations consisting of two diseases.

Since m_6 is absent, the final solution to the problem is given by these same two generators (last line in Figure 4). Note that these two generators implicitly represent the eight explanations for M^+ that were listed earlier. It is also interesting to note that d_3 and d_4, eliminated from the FOCUS when m_4 was found to be present, are once again viable possibilities. Had this been a larger knowledge base with additional manifestations, the FOCUS would have continued to evolve using similar set intersection operations.

4. An Example Application of the Model

In the interest of clarity, we have so far ignored several aspects of real-world diagnostic problem-solving in presenting the set cover model. Rather than immediately examining these details, we will first briefly illustrate an interactive session with System D.

Expert System D must handle situations where multiple causes of dizziness are present simultaneously. The following conversation illustrates just such a situation. User typing is underlined and ellipsis indicates where material has been excised for brevity. The system's tracing mechanism has been turned on producing the normally unseen output highlighted in boxes which represents periodic "snapshots" of the FOCUS.

.
.
.

DIZZINESS DIAGNOSIS DECISION SUPPORT SYSTEM
TYPE 'CONTINUE' TO BEGIN

READY FOR COMMAND:
assert dizziness.

(Rather than waiting to be asked questions, the user elects to volunteer some initial information.)

FOR DIZZINESS = PRESENT

TYPE:
   (1) DEFINITE ROTATIONAL SENSATION
   (2) SENSATION OF IMPENDING FAINT
   (3) ILL-DEFINED LIGHT-HEADEDNESS
   (4) LOSS OF BALANCE WITHOUT HEAD SENSATION

= ?2/3.
COURSE:
   (1) ACUTE OR SUBACUTE PERSISTENT DEFICIT LASTING MORE THAN ONE
      HOUR
   (2) ONE OR MORE CLEAR-CUT EPISODES
   (3) GRADUAL ONSET OR SLOW PROGRESSION WITHOUT CLEAR-CUT
EPISODES
= ?2.

(The user indicates that the patient has episodic dizziness described as faintness or light-headedness. The answer "2/3" means "2 *or* 3".)

      .
      .
      .

READY FOR COMMAND:
assert blood pressure examination = significant
   orthostatic hypotension.

(The user indicates that the patient's blood pressure falls when he stands up, i.e., there is orthostatic hypotension.)

READY FOR COMMAND:
obtain diagnosis.

(The user activates System D's inference mechanism, which is based on the set covering process explained above.)

FOCUS:
  GENERATOR:
    MULTIPLE SYSTEM ATROPHY
    IDIOPATHIC ORTHOSTATIC HYPOTENSION
    SYMPATHETIC ORTHOSTATIC HYPOTENSION
    AUTONOMIC NEUROPATHY
    ORTHOSTATIC HYPOTENSION SECONDARY TO PHENOTHIAZINES
    ORTHOSTATIC HYPOTENSION SECONDARY TO ANTIDEPRESSANTS
    ORTHOSTATIC HYPOTENSION SECONDARY TO L-DOPA
    ORTHOSTATIC HYPOTENSION SECONDARY TO ANTIHYPERTENSIVE
      MEDICATIONS
    ORTHOSTATIC HYPOTENSION SECONDARY TO DIURETICS
    HYPERBRADYKINISM
    ORTHOSTATIC HYPOTENSION SECONDARY TO PARKINSONISM
    ORTHOSTATIC HYPOTENSION SECONDARY TO PREVIOUS SYMPA-
      THECTOMY
    ORTHOSTATIC HYPOTENSION SECONDARY TO PROLONGED RE-
      CUMBENCY

(The initial FOCUS for the two known manifestations, dizziness and ortho-static hypotension, consists of a single generator which in turn consists of a single set of competing diseases. Each disease represents a minimal cover for the two manifestations.)

NEURO-OTOLOGICAL SYMPTOMS ASSOCIATED WITH DIZZINESS:
  (1) DIPLOPIA
  (2) LOSS OF OR BLURRED VISION
  (3) SCINTILLATING SCOTOMAS
  (4) FOCAL SYMPTOMS REFERRABLE TO CNS
  (5) SYNCOPE
  (6) HEARING LOSS BY HISTORY
  (7) TINNITUS
= ?3&5.

(Based on the active hypothesis, System D generates its first hypothesis-driven question and discovers two new manifestations, syncope and scintillating scotomas.)

FOCUS:
  GENERATOR:
    BASILAR MIGRAINE
      &
    ORTHOSTATIC HYPOTENSION SECONDARY TO PROLONGED RE-
      CUMBENCY
    ORTHOSTATIC HYPOTENSION SECONDARY TO SYMPATHECTOMY
    ORTHOSTATIC HYPOTENSION SECONDARY TO PARKINSONISM
    HYPERBRADYKINISM
    ORTHOSTATIC HYPOTENSION SECONDARY TO DIURETICS
    ORTHOSTATIC HYPOTENSION SECONDARY TO ANTIHYPERTENSIVE
      MEDICATIONS
    ORTHOSTATIC HYPOTENSION SECONDARY TO L-DOPA
    ORTHOSTATIC HYPOTENSION SECONDARY TO ANTIDEPRESSANTS
    ORTHOSTATIC HYPOTENSION SECONDARY TO PHENOTHIAZINES
    AUTONOMIC NEUROPATHY
    SYMPATHOTONIC ORTHOSTATIC HYPOTENSION
    IDIOPATHIC ORTHOSTATIC HYPOTENSION
    MULTIPLE ATROPHY

(None of the previous possible explanations of the patient's manifestations can now account for the new manifestations. In particular, SCINTILLAT-ING SCOTOMAS is not explained. A new FOCUS is therefore created consisting of a single generator representing minimal covers containing two disorders. The "&" here represents the "×" used in generators as described in section 3.)

CURRENT MEDICATIONS:
  (1) ANTICOAGULANTS
  (2) LARGE AMOUNTS OF QUININE
  (3) PHENOTHIAZINES
  (4) ANTIHYPERTENSIVE AGENTS
    . . .

(12) BARBITURATES
= ?2&3&4.

```
ELEMENTARY DISORDERS NOW CATEGORICALLY REJECTED:
  DIZZINESS SECONDARY TO BARBITURATES
  OTOTOXICITY SECONDARY TO AMNIOGLYCOSIDES
  OTOTOXICITY SECONDARY TO SALICYLATES
  ORTHOSTATIC HYPOTENSION SECONDARY TO ANTIDEPRESSANTS
  ORTHOSTATIC HYPOTENSION SECONDARY TO L-DOPA
  ORTHOSTATIC HYPOTENSION SECONDARY TO DIURETICS
```

(After learning the patient's medications, System D is able to narrow down the number of possibilities. How this is done is described in the next section.)

ABNORMALITIES ON NEUROLOGICAL EXAMINATION:
   (1) OPTIC ATROPHY
   (2) PAPILLEDEMA
   (3) HOMONYMOUS FIELD CUT
   (4) NYSTAGMUS
   (5) IMPAIRED HEARING
   (6) PERIPHERAL NEUROPATHY
   (7) PARKINSONISM
   (8) FOCAL CNS FINDINGS
= ?3&4&5.

(The user indicates the presence of three additional manifestations: homonymous hemianopsia, nystagmus, and impaired hearing.)

```
FOCUS:
  GENERATOR:
    BASILAR MIGRAINE
      &
    COGAN'S SYNDROME
    OTOTOXICITY SECONDARY TO QUININE
    OTOSCLEROSIS
    LABRINTHINE FISTULA
    MENIERE'S DISEASE
      &
    ORTHOSTATIC HYPOTENSION SECONDARY TO PROLONGED RE-
      CUMBENCY
    ORTHOSTATIC HYPOTENSION SECONDARY TO PREVIOUS SYMPA-
      THECTOMY
    ORTHOSTATIC HYPOTENSION SECONDARY TO PARKINSONISM
    HYPERBRADYKINISM
    ORTHOSTATIC HYPOTENSION SECONDARY TO ANTIHYPERTENSIVE
      MEDICATIONS
    ORTHOSTATIC HYPOTENSION SECONDARY TO PHENOTHIAZINES
    AUTONOMIC NEUROPATHY
    IDIOPATHIC ORTHOSTATIC HYPOTENSION
    MULTIPLE SYSTEM ATROPHY
```

(The new FOCUS resulting from the additional information is illustrated here. Each explanation now consists of three disorders. The previous explanations could not account for the impaired hearing or nystagmus, and these new manifestations are now assumed to be due to one of five causes: COGAN'S SYNDROME, etc. Note that this FOCUS represents 1 * 5 * 9 = 45 potential explanations in a compact fashion.)

READY FOR COMMAND:
display value (diagnosis).

    BASILAR MIGRAINE ⟨A⟩
      &
    OTOTOXICITY SECONDARY TO QUININE ⟨H⟩
    OTOSCLEROSIS ⟨M⟩
    LABYRINTHINE FISTULA ⟨L⟩
    MENIERE'S DISEASE ⟨L⟩
      &
    ORTHOSTATIC HYPOTENSION SECONDARY TO ANTIHYPERTENSIVE MEDI-
      CATIONS ⟨H⟩
    ORTHOSTATIC HYPOTENSION SECONDARY TO PHENOTHIAZINES ⟨H⟩
    IDIOPATHIC ORTHOSTATIC HYPOTENSION ⟨M⟩
    AUTONOMIC NEUROPATHY ⟨M⟩
    MULTIPLE SYSTEM ATROPHY ⟨L⟩
    ORTHOSTATIC HYPOTENSION SECONDARY TO PARKINSONISM ⟨L⟩

    READY FOR COMMAND:
      .   .   .

This final diagnosis offered by System D, including a ranking of competing alternatives which will be explained below, means: "The patient has basilar migraine. In addition, the patient also probably has ototoxicity secondary to the quinine he is taking, although he could have otosclerosis or even one of the other unlikely inner ear disorders listed. Finally, the patient also has orthostatic hypotension which is probably due to his medications, but might be due to one of the other listed causes." This final diagnostic account of the patient's complex set of signs and symptoms is very plausible.

5. From Model to Functioning Expert System

As noted earlier, the implementation of functioning expert systems like System D based on the set covering model requires that several issues be addressed and resolved. We have already discussed adopting the model to sequential problem solving, so we now turn to several other aspects of real-world diagnostic problem-solving. Further details about these issues can be found in the references (Reggia, 1981).

Question Generation and Termination Criteria

The vast majority of questions generated by Expert System D, representing Step 1 in the informal sequential algorithm presented earlier, fall into the category of hypothesis-driven questions. In other words, each question is based solely on the disorders in the FOCUS at that point during problem-solving. Let us say that a disorder is *active* if it is currently in the FOCUS. Then to select its next question the expert system extracts from the DESCRIPTION of each active disorder the first attribute in an assertion whose current value is not yet known (recall that assertions, attributes and values were defined in section 3, "Case-Specific Database"). From these candidate attributes, the one in the largest number of DESCRIPTIONs of active disorders is selected to form the basis of the next question.

This simple, heuristic approach to question generation makes no claim to optimality. However, it does have certain properties that make it a useful strategy to follow. Since it selects one of the most commonly referred to attributes of active disorders, it usually produces questions that help to discriminate among the competing explanations in the FOCUS. In addition, since it selects candidate questions from the *first* unknown attributes remaining in these DESCRIPTIONs, it allows the knowledge base author to exert partial control over the order in which questions are generated (i.e., by consistently ordering the assertions in DESCRIPTIONs in a similar fashion). Finally, this approach to question generation has the advantage of being computationally inexpensive when compared with more elaborate optimization schemes that might be used.

Once a new question has been asked and answered by the user, another hypothesize-and-test cycle begins. This continues until no further questions can be generated because no assertions in the DESCRIPTIONs of active disorders contain attributes whose values have not been acquired from the user. This termination condition is a somewhat arbitrary approach to deciding when sufficient information is known. While it asks about all attributes relevant to ranking the competing explanations involved at termination time, it might leave some information unsought. To permit the knowledge base author to insure the level of completeness of information collection that is desired from an expert system, protocol-driven questions that should always be asked may optionally be included as explicit instructions to an expert system at the time it is constructed.

Setting Factors and the Ranking of Competing Disorders

Once the termination condition is satisfied, expert systems like System D enter a final scoring phase during which competing disorders are ranked relative to one another for the first time. In other words, the hypothesize-and-test control cycles have previously only been concerned with the construction of all possible explanations (a differential diagnosis) without regard to their relative likelihood. For each active disorder at termination time two numeric scores are calculated: a *setting score* and a *match score*. These scores are calculated using the symbolic probabilities in the knowledge base as well as any symbolic probabilities incorporated in a user's response to questions. A simple weighting scheme ($A = 4$, $H = 3, \ldots$, $N = 0$) is used for these calculations.

The setting score for an active disorder is initialized to the numerical equivalent of its symbolic probability originally specified following its name in the knowledge base. This initial score is then incrementally adjusted upwards or downwards based only on assertions about setting factors in its DESCRIPTION. The setting score is intended to provide a generalization of the concept of prior probability in that it reflects the general likelihood of a disorder in the context of the specific setting in which it is occurring.

The match score of an active disorder is based only on M^+ and the assertions about manifestations in its DESCRIPTION. The match score is also derived using a simple weighting scheme. At termination time, an expert system conceptually has in the FOCUS all of the possible competing explanations for M^+. It can therefore derive a match score for any disorder based on the "best" explanation which contains that disorder (i.e., the explanation that as a whole would be most likely to cause M^+). For example, if d_1 is in two explanations, then the match score for d_1 would be based on its role in the "better" of these two possible explanations. Furthermore, a manifestation can be assigned to the disorder in an explanation which is most likely to be producing it in situations where that manifestation can be caused by more than one of the disorders in the explanation. The match score is intended as a measure of how closely a disorder fits the manifestations of a case, irrespective of the setting in which they are occurring.

A final score is calculated for each active disorder based on both its setting score and its match score. Since this final numerical weighting is intended to provide only a "ballpark" indication of how likely the disorder is, it is subsequently converted back into a symbolic probability to emphasize its imprecise and heuristic nature. This was illustrated in the conversation with System D when the final diagnostic possibilities were listed in order of likelihood.

It should be appreciated that the set covering model used in this fashion permits scoring which can be considered to be truly context-dependent. Not only does a disorder's likelihood depend on the specific environment in which it is occurring (setting score), but it also depends on what other disorders are postulated to be simultaneously present in an explanation, and on which of several competing explanations contain it (match score).

One final point about the use of symbolic probabilities needs to be made. While the ranking of competing disorders is done *after* the termination condition is satisfied, the symbolic probabilities A and N are used in one other way *during* the hypothesize-and-test control cycles. They are used to determine when any disorder d_i should be *categorically rejected* by the inference mechanism. For example, the DESCRIPTION of ORTHOSTATIC HYPOTENSION SECONDARY TO L-DOPA in System D's knowledge base contains the categorical assertion

CURRENT MEDICATIONS = L-DOPA ⟨A⟩.

Thus, if System D discovered that a patient was not taking L-DOPA, ORTHOSTATIC HYPOTENSION SECONDARY TO L-DOPA would be immediately discarded from any further consideration by the inference mechanism (as occurred after the question on CURRENT MEDICATIONS in the conversation with System D earlier). In effect, what occurs is that the set D is changed: the set of all possible disorders is modified by removing any disorders discovered to be categorically rejected during problem-solving. All subsequent development of the SCOPE and FOCUS by the inference mechanism reflects this change in the very framework of the problem.

Problem Decomposition

Since finding a minimal set cover is known to be NP complete (Karp, 1972), the task of constructing the solution to a diagnostic problem is potentially combinatorially expensive as the size of an explanation increases. This difficulty is only academic for some classes of diagnostic problems. For example, it is not uncommon for a patient seen by a physician to have more than one disease simultaneously, but it would be exceedingly rare for someone to have more than 50 diseases simultaneously. However, since the potential for combinatorial explosion exists, it is still important to address the question of when a diagnostic problem can be reduced or decomposed into smaller, independent subproblems.

One example of when this can be done is best presented by introducing the concept of "connected" manifestations. Two manifestations m_a and m_b are said to be *connected* if either causes(m_a) and causes(m_b) have a non-empty intersection, or there exists a finite set of manifestations $\{m_1, m_2, \ldots, m_n\}$ such that $m_1 = m_a$, $m_n = m_b$, and each m_j is connected to m_{j+1}. All of the manifestations appearing in Table 1, for example, are connected to one another. It can be shown that if M^+ can be partitioned into N subsets of connected manifestations, each subset of which contains no manifestation connected to another manifestation in a different subset, then the original diagnostic problem can be partitioned into N independent subproblems. The generators for the solution to the original problem are then easily constructed by appending in an appropriate fashion the generators for the solutions to the subproblems (Reggia, 1981).

Furthermore, sequentially constructing and maintaining independent subproblems in this way, each with its own SCOPE, FOCUS and MANIFS, is relatively easy. When a new manifestation m_i is found to be present, the set causes(m_i) is intersected with the SCOPE of each pre-existing subproblem. When this intersection is non-empty, m_i is said to be *related* to the corresponding subproblem. There are three possible results of identifying the subproblems to which m_i is related. First, m_i may not be related to any pre-existing subproblems. In this case, a new subproblem is created, with MANIFS = $\{m_i\}$, SCOPE = causes(m_i), and FOCUS = a single generator consisting of the single set of competing disorders found in causes(m_i). This is what always occurs when the first manifestation becomes known, as was illustrated in Figure 4. Second, m_i may be related to exactly one subproblem, in which case m_i is assimilated into that subproblem as described earlier and illustrated with m_4 and m_5 in Figure 4. Finally, m_i may be related to multiple existing subproblems. In this situation, these subproblems are "joined" together to form a new subproblem, and m_i is then assimilated into this new subproblem (not illustrated in Figure 4 nor in the conversation with System D, both of which involved only a single subproblem).

Other Considerations

Many other considerations go into expanding the generality and robustness of the set covering model for use in real world expert systems. We will mention just three of these issues here: unexplainable manifestations, the single-disorder constraint, and non-independent disorders.

Assuming that an expert system's knowledge base and a relevant case are both correct and complete, the set cover model as described above can handle a broad range of diagnostic problems. Unfortunately, in the real world, this ideal situation is sometimes not present. A knowledge base might be incomplete or contain errors, especially during system development, and a user might enter incorrect information about a problem.

One example of such a situation is the *unexplainable manifestation:* a manifestation m_j whose associated set causes(m_j) is empty. If undetected, such a manifestation would result in repeated futile attempts by the inference mechanism to create progressively larger and larger explanations to account for all known manifestations. The important point here is that the inference mechanism must continuously monitor for unexplainable manifestations at run time. This is because an initially non-empty set causes(m_j) could potentially become empty during problem-solving if all of the disorders in it were discovered to be categorically rejected. Our expert systems currently handle this anomaly by informing the user of the situation, discarding the unexplainable manifestation, and offering the user the option of continuing with the understanding that all is not well.

Another issue deserving special attention is situations where only one disorder is expected to occur at a time. Even though such a *single-disorder constraint* may not be strictly correct in a theoretical sense, there are situations where such an assumption is justified by practical considerations. An example of an expert system called System P which uses the set covering model with the single disorder constraint is given in Appendix B. System P uses this constraint because of the exceedingly low likelihood that two of the individually very rare disorders in its knowledge base would occur in a single individual. The advantage of using the single-disorder constraint when appropriate is that it permits the automatic recognition by the inference mechanism of potential errors. This is illustrated in the conversation with System P in Appendix B when it indicates that no single disorder can account for all of the facts in the case under consideration. Such a situation might have been due to (i) user error in describing the case, (ii) an incomplete or incorrect knowledge base, or (iii) a patient with a previously unknown form of peroneal muscular atrophy.

When the single-disorder constraint is employed, two adjustments are made to the inference mechanism of expert systems using the set covering model. First, at the start of a case the FOCUS is initialized to a single generator whose single set includes all possible disorders within the domain of the expert system. This initial FOCUS represents the initial hypothesis that exactly one possible disorder is present. Second, the inference mechanism as usual monitors for the occurrence of an empty FOCUS, but interprets such an occurrence as an anomaly. It does not try to construct explanations containing two disorders, but indicates to the user that it cannot explain the current case findings with a single disorder (see conversation with System P).

Finally, we have assumed so far that the disorders in D are independent of one another, an assumption that may not be valid in some domains. One possible approach to this non-independence would be to award a "bonus" during scoring to explanations where associated disorders were involved [this was the approach used in INTERNIST; see Pople, Myers & Miller (1975)]. We have elected to study instead those situations where disorders can be partitioned into classes, with disorders in one class causing disorders in another. For example, one expert system currently being constructed involves both localization of damage in the nervous system and diagnosis.

6. Discussion

This paper has proposed the construction and maintenance of minimal set covers ("explanations") as a general model of diagnostic reasoning and has illustrated its use as an inference method for diagnostic expert systems. The set cover model is attractive in that it directly handles multiple simultaneous disorders, it can be formalized, it is intuitively plausible, and it is justifiable in terms of past empirical studies of diagnostic reasoning (e.g., Elstein *et al.,* 1978; Kassirer & Gorry, 1978). To our knowledge the analogy between the classic set covering problem and general diagnostic reasoning has not previously been examined in detail, although some related work has been done [for example, assignment of HLA specificities to antisera; Nau, Markowsky, Woodbury & Amos (1978) and Woodbury, Ciftan & Amos (1979)].

The set cover model provides a useful context in which to view past work on diagnostic expert systems. In contrast to the set cover model, most diagnostic expert systems that use hypothesize-and-test inference mechanisms or which might reasonably be considered as models of human diagnostic reasoning depend heavily upon the use of production rules (e.g., Aikins, 1979: Mittal, Chandrasekaran & Smith, 1979; Pauker, Gorry, Kassirer & Schwartz, 1976). These systems use a hypothesis-driven approach to guide the invocation of rules which in turn modify the hypothesis. A rule-based hypothesize-and-test process does not provide a convincing model of what has been learned about human diagnostic reasoning in the empirical studies cited earlier. Furthermore, rules have long been criticized as a representation of diagnostic knowledge (e.g., Reggia, 1978), and their invocation to make deductions or perform actions does not capture in a general sense such intuitively attractive concepts as coverage, minimality, or explanation.

Perhaps the previous diagnostic expert system whose inference method is closest to the set cover model is INTERNIST (Pople *et al.,* 1975). INTERNIST is a large and well-known expert system that represents diagnostic knowledge in a DESCRIPTION-like fashion and does not rely on production rules to guide its hypothesize-and-test process. In contrast to the set cover model, however, INTERNIST's inference mechanism uses a heuristic scoring procedure to guide the construction and modification of its hypothesis. This process is essentially serial or depth-first, unlike the more parallel or breadth-first approach implied in the set cover model. In other words, INTERNIST first tries to establish one disorder and then proceeds to establish others. This roughly corresponds to constructing and completing a single generator set in the set cover model, and then later returning to construct the additional sets for the generator. The criteria used by INTERNIST to group together competing disorders (i.e., a set in a generator) is based on a simple heuristic: "Two diseases are competitors if the items not explained by one disease are a subset of the items not explained by the other; otherwise, they are alternates (and may possibly coexist in the patient)" (Miller, Pople & Myers, 1982). In the terms of our model, this corresponds to stating that d_1 and d_2 are competitors if $M^+ - man(d_1)$ contains or is contained in $M^+ - man(d_2)$. It can be proven that while this simple heuristic may generally work in constructing a differential diagnosis, there are clearly situations for which it will fail to correctly group competing disorders together.† Reportedly, the serial or depth-first approach used in INTERNIST resulted in less than optimal performance (Pople, 1977; Miller *et al.,* 1982), and it has been criticized as *"ad hoc"* by some individuals working in statistical pattern classification because of the lack of a formal underlying model (e.g., Ben-Bassat *et al.,* 1980). It is also unclear that the INTERNIST inference mechanism is guaranteed to find all possible explanations for a set of manifestations. Recent enhancements in INTERNIST's successor CAD-UCEUS attempt to overcome some of these limitations through the use of "constrictors" to delineate the top-level structure of a problem (Pople, 1977). These changes are quite distinct from the approach taken in the set cover model, but do add a breadth-first component to hypothesis construction.

The set cover model presented here is still evolving both theoretically and in terms of its evaluation in practice. Work is clearly needed in at least three directions: further theoretical development of the model, assessment of its application in expert systems involving a broad range of real-world diagnostic problems, and assessment of its adequacy as a cognitive model. We intend to pursue these issues in the future.

Acknowledgment. This research was supported by NIH grants 5 K07 NS 00348 and 1 P01 NS 16332 and in part by NSF grant MCS81-17391. Computer time was provided by the Computer Science Center of the University of Maryland.

Appendix A: Table of Symbols

Symbol	Meaning
$\in$	element of
$\subseteq$	subset of
$\leq$	less than or equal to
$\emptyset$	empty set
$\forall$	for all
$\cup$	set union
$\cap$	set intersection
$\mid \ \mid$	set cardinality
M	universe of manifestations (effects)
D	universe of disorders (causes)
C	causative relationship

†For example, suppose $M^+ = \{m_1 \cdots m_8\}$ and only d_1, d_2, and d_3 have been evoked where $M^+ \cap$ man $(d_1) = \{m_2\ m_4\ m_5\ m_6\ m_7\ m_8\}$, $M^+ \cap$ man$(d_2) = \{m_3\ m_4\ m_5\ m_6\ m_7\ m_8\}$, and $M^+ \cap$ man$(d_3) = \{m_1\ m_2\ m_3\}$. In the set cover model, Sol(P) $= \{\{d_1\ d_3\}\{d_2\ d_3\}\}$ which can be represented by the single generator $\{d_1\ d_2\} \times \{d_3\}$ where d_1 and d_2 are grouped together as competitors. Suppose that d_1 was ranked highest by the INTERNIST heuristic scoring procedure. Then $M^+ -$ man$(d_1) = \{m_1\ m_3\}$ and $M^+ -$ man$(d_2) = \{m_1\ m_2\}$, so INTERNIST would apparently fail to group d_1 and d_2 together as competitors.

Appendix B

Expert System P addresses the problem of diagnosing the cause of wasting of the muscles of the lower legs (peroneal muscular atrophy). The causes of this problem are very rare, and include entities such as hypertrophic neuropathy of Charcot-Marie-Tooth and Refsum's Syndrome. Because of the rarity of each of these disorders, it is extremely unlikely that more than one would occur in an individual, so System P uses the set covering model with the additional constraint that only one possible cause can be present in any case. This leads to the following interesting result (user typing is underlined in the conversation, and the system's tracing mechanism is turned on producing the output highlighted in boxes which represent normally unseen periodic "snapshots" of the FOCUS):

> .
> .
> .

PERONEAL MUSCULAR ATROPHY DECISION SUPPORT SYSTEM FOR DIAGNOS-
TIC CLASSIFICATION. TYPE 'CONTINUE' TO BEGIN.

READY FOR COMMAND:
continue
OK—PLEASE ANSWER THE FOLLOWING . . .
AGE OF ONSET:
   (1) INFANCY
   (2) CHILDHOOD OR ADOLESCENCE
   (3) ADULT LIFE
= ?2.

(The manner in which the system selects questions to ask was addressed in section 5.)

FOCUS:
GENERATOR:
   HEREDITARY SPASTIC PARAPLEGIA WITH PERONEAL MUSCULAR
     ATROPHY
   FRIEDRICH'S ATAXIA
   REFSUM'S SYNDROME
   PROGRESSIVE SPINAL MUSCULAR ATROPHY OF CHARCOT-MARIE-
     TOOTH
   NEURONAL TYPE OF CHARCOT-MARIE-TOOTH DISEASE
   NON-DOMINANT HYPERTROPHIC NEUROPATHY OF CHARCOT-MARIE-
     TOOTH
   HYPERTROPHIC NEUROPATHY OF CHARCOT-MARIE-TOOTH

(Knowing that exactly one cause of peroneal muscular atrophy is present, System P starts off with an initial hypothesis consisting of a single generator containing the eight possibilities. One cause is immediately discarded because it always starts in infancy, leaving the seven potential explanations listed here.)

INHERITANCE:
   (1) SPORADIC
   (2) RECESSIVE

    (3) DOMINANT
= ?unknown.
SEVERITY OF MOTOR DEFICIT:
    (1) MILD
    (2) MODERATE
    (3) SEVERE
= ?3.

```
FOCUS:
    GENERATOR:
        FRIEDRICH'S ATAXIA
        REFSUM'S SYNDROME
```

(The three possible causes of a severe motor deficit are FRIEDRICH'S ATAXIA, REFSUM'S SYNDROME, and HYPERTROPHIC NEUROP-ATHY OF DEJERINE-SOTTAS. Intersecting this set with the previous generator decreases the alternative explanations to the two listed here.)

SENSORY DEFICIT:
    (1) PRESENT
    (2) ABSENT
= ?1.

FOR SENSORY DEFICIT =
    PRESENT . . .

    SEVERITY:
    (1) MILD
    (2) MODERATE
    (3) SEVERE
= ?2.

(When told that the patient has a sensory deficit, System P automatically asks for relevant elaboration about its severity. This request for elaboration is the only example of a protocol-driven question in this conversation. The discovery of a sensory deficit does not change the FOCUS, shown imme-diately below, because this manifestation can be caused by both active disorders.)

```
FOCUS:
    GENERATOR:
        FRIEDRICH'S ATAXIA
        REFSUM'S SYNDROME
```

PES CAVUS:
    (1) PRESENT
    (2) ABSENT
= ?1.

```
FOCUS:
    GENERATOR:
        FRIEDRICH'S ATAXIA
```

(A third manifestation, PES CAVUS or "high arches," is uncovered, and intersecting its set of causes with the FOCUS results in a single remaining possible explanation. Rather than accepting this hypothesis as the diagnosis, System P continues to generate questions to confirm it.)

```
OTHER NEUROLOGICAL FINDINGS:
   (1)  RETINITIS PIGMENTOSA
   (2)  DEAFNESS
   (3)  NYSTAGMUS
   (4)  DYSARTHRIA
   (5)  CEREBELLAR ATAXIA
   (6)  PYRAMIDAL SIGNS
= ? (MULTIPLE ANSWERS PERMITTED)
1.

A SINGLE VALUE OF
TYPE OF PERONEAL MUSCULAR ATROPHY CANNOT EXPLAIN
ALL OF THE FEATURES OF THIS CASE
SHOULD PROCESSING CONTINUE USING THOSE VALUES THAT
ARE NOT CATEGORICALLY REJECTED
= ? (YES/NO)
no.

TYPE OF PERONEAL MUSCULAR ATROPHY =
   UNKNOWN

READY FOR COMMAND: . . . .
```

(When System P learns that RETINITIS PIGMENTOSA is present, intersection of its causes with the generator results in an empty FOCUS. System P gives up and classifies this patient as having an unknown disease.)

What is striking here is that System P automatically detects that it does not know the diagnosis in this case. This is because the special constraint that only a single disorder be present, imposed by the creator of System P, contradicts the definition of adequacy required of an explanation in the context of this specific case.

References

Aikins, J. (1979). Prototypes and production rules: an approach to knowledge representation for hypothesis formation. *Proceedings of the Sixth International Joint Conference on Artificial Intelligence,* Tokyo, Japan, pp. 1–3.

Ben-Bassat, M., Carlson, R., Puri, V., Davenport, M., Schriver, J., Latif, M., Smith, R., Portigal, L., Lipnick, E. & Weil, M. (1980). Pattern-based interactive diagnosis of multiple disorders: the MEDAS system. *IEEE Transactions on Pattern Analysis and Machine Intelligence,* **2**, 148–160.

Edwards, J. (1962). Covers and packings in a family of sets. *Bulletin of the American Mathematics Society,* **68**, 494–499.

Elstein, A., Shulman, L. & Sprafka, S. (1978). *Medical Problem Solving—An Analysis of Clinical Reasoning.* Cambridge, Massachusetts: Harvard University Press.

Karp, R. (1972). Reducibility among combinatorial problems. In Miller, R. & Thatcher, J., Eds, *Complexity of Computer Computations,* pp. 85–103. New York: Plenum Press.

Kassirer, J. & Gorry, G. (1978). Clinical problem solving: a behavioral analysis. *Annals of Internal Medicine,* **89,** 245–255.

Miller, R., Pople, H. & Myers, J. (1982). INTERNIST-1, An experimental computer-based diagnostic consultant for general internal medicine. *New England Journal of Medicine,* **307,** 468–476.

Mittal, S., Chandrasekaran, B. & Smith, J. (1979). Overview of MDX: a system for medical diagnosis. *Proceedings of the Third Annual Symposium on Computer Applications in Medical Care,* pp. 34–46. Piscataway, New Jersey: IEEE Press.

Nau, D., Markowsky, G., Woodbury, M. & Amos, D. (1978). A mathematical analysis of human leukocyte antigen serology. *Mathematical Biosciences,* **40,** 243–270.

Pauker, S., Gorry, G., Kassirer, J. & Schwartz, W. (1976). Towards the simulation of clinical cognition. *American Journal of Medicine,* **60,** 981–996.

Pople, H. (1977). The formation of composite hypotheses in diagnostic problem solving: an exercise in synthetic reasoning. *Proceedings of the Fifth International Joint Conference on Artificial Intelligence,* Carnegie-Mellon University, Pittsburgh, Pennsylvania, pp. 1030–1037.

Pople, H., Myers, J. & Miller, R. (1975). DIALOG: a model of diagnostic logic for internal medicine. *Proceedings of the Fourth International Joint Conference on Artificial Intelligence,* pp. 848–855. Cambridge, Massachusetts: M.I.T. Artificial Intelligence Laboratory Publications.

Reggia, J. (1978). A production rule system for neurological localization. *Proceedings of the Second Annual Symposium on Computer Applications in Medical Care,* pp. 254–260. Piscataway, New Jersey: IEEE Press.

Reggia, J. (1981). Knowledge-based decision support systems: development through KMS. *TR-1121,* Department of Computer Science, University of Maryland, College Park.

Reggia, J. (1982). Computer-assisted medical decision making. In Schwartz, M., Ed., *Applications of Computers in Medicine,* pp. 198–213. Piscataway, New Jersey: IEEE Press.

Woodbury, M., Ciftan, E. & Amos, D. (1979). HLA serum screening based on an heuristic solution of the set cover problem. *Computer Programs in Biomedicine,* **9,** 263–273.

VI. RELATED ISSUES

The Evaluation of Clinical Predictions: A Method and Initial Application

Alan R. Shapiro

Clinical predictions are never certain but are inherently probabilistic. The accuracy coefficient, a measure of probabilistic accuracy based on probability assigned to outcomes that occur, was used to assess the skill of clinical rheumatologists in predicting patient outcomes. Physicians' scores correlated well with degree of clinical experience. An approach to evaluation based on the measure provides a sensitive assessment of marginal benefit of technologies such as laboratory tests, diagnostic procedures or computer consultations. Most currently used methods of computer prediction were not as accurate as the best physicians tested. By allowing measurement of ability to individualize predictions to each patient's unique characteristics, the accuracy-coefficient approach has potential use in physician assessment. (N Engl J Med 296:1509–1514, 1977)

Predictive thinking plays an essential part in clinical medicine. The ability to predict the future underlies medical practice and is basic to decisions concerning diagnosis, prognosis and therapy. But experience with clinical prediction from the time of Hippocrates's *Book of Prognostics* has shown that predictions can rarely be certain and need, therefore, to be qualified in probabilistic terms. In deciding whether to add a potentially toxic agent to a patient's regimen, a physician must consider, at least implicitly, the probability that the patient will improve or deteriorate without added therapy, the chances of improvement if the therapy is added and the risks of different types of toxicity due to therapy. Proper medical decisions must be based on accurate probabilistic predictions. The improvement of the physician's ability to make these predictions can be recognized as a primary motivating force behind medical research. Yet attempts in medicine to analyze physicians' ability at prediction have been rare.

Reprinted with permission from the *New England Journal of Medicine, 296*, pp. 1509–1514, 1977.

The need for a method to evaluate probabilistic predictions has recently attained a certain urgency. Computerized procedures have been introduced in medicine for providing "exact" probabilistic predictions of outcomes such as failure to recover from coma [1,2], favorable response to coronary surgery [3,4] and deterioration of renal function in systemic lupus erythematosus [5]. Clearly, momentous decisions affecting the life of a patient may be made if a physician learns on the basis of such prediction procedures that a patient has a very high probability of incurring any such outcome. It is important to realize that such predictions can be wrong.

This article introduces a method for assessing the predictive skills of physicians and presents the results of recent research on clinical prediction. In addition, the article outlines potential uses of the method in evaluating the contribution of such technologies as computer consultations or diagnostic procedures to the predictive accuracy of clinicians.

Critique of Existing Methods of Evaluating Predictions

It may seem difficult to imagine how a prediction couched in probabilistic terms can be considered wrong. If a physician or a computerized predictor asserts that the chance of death from a certain operation is only 5 per cent and the patient then submits to the operation and dies, it cannot be concluded that the physician was wrong. The prediction said that a favorable outcome was highly probable—one may claim that is was probable but unfortunately an improbable event occurred. Although such reasoning shows that probabilistic statements cannot be disproved, particularly in the individual case, it is still reasonable to attempt the objective evaluation of groups of predictions.

The simplest and most widely used method of evaluation is to examine the proportion of predictions in which an error is made. If a physician makes the prediction that it is more likely than not that a patient will be cured by operation and the patient then is not cured, the physician has, in retrospect, made an error. The proportion of such errors is the error rate. Concepts such as sensitivity and specificity, which are used in the evaluation of diagnostic procedures, are based on the error-rate method. This method is appealing because of its simplicity, but there are two reasons why it is inappropriate for examining physician predictions.

The first is that the error rate is insensitive. It does not take into account the magnitude of an error. Using the error-rate method to evaluate the predictive accuracy of a physician or of a laboratory test is similar to evaluating the accuracy of two thermometers solely on the basis of their ability to distinguish febrile from afebrile patients. In predicting whether an operation will be successful or not, as long as the probability of success was said to be more than 0.50 and the operation was successful, the prediction will be regarded as correct by the error-rate method. The prediction will receive the same score regardless of whether the probability of success was realistically stated to be slightly more than 0.50 or optimistically asserted to be 99,999 in 100,000. The method does not penalize the physician who chooses to forecast dire outcomes and "hang crepe" rather than attempt accurate predictions [6].

The second reason is that the use of error rates to assess predictions fails to distinguish a decision from the evidence on which that decision is based. To be sure, from the standpoint of the patient, the over-riding concern is the correctness of a decision. From the perspective of the evaluation of predictive ability, however, the final decision is the result of several components. These components include value assessments such as the meaning to an individual patient of relief from symptoms, of an adverse drug reaction, of costly bills for services and of many other outcomes whose meaning is unique to a particular case, and predictive assessments concerning the probability of such events. Examination of these several predictions provides a deeper insight into the physician's cognitive process than is obtained by the sole observation of the success of the final decision. Decisions can be considered right or wrong; probabilities are never right or wrong but are best considered more or less accurate. What is needed in assessing probabilistic predictions is a method for assessing the accuracy of the probabilities asserted, not one designed for scoring the all-or-none rectitude of decisions. Efforts toward this end have been reported in meteorology, a field in which the probabilistic basis of prediction has been explicitly recognized [7,8].

A Measure of Predictive Accuracy

A measure for evaluation of predictive accuracy is presented below.

Consider the probabilities assigned by two different physicians to a series of clinical events in which there are two possible outcomes: the event either will or will not take place. If one physician states that the outcomes that occur were unlikely to have happened, and another physician assigns a much higher probability to these actual outcomes then, with respect to predictive skill, one would tend to favor the physician who assigned the higher probability to the outcomes that actually occurred. It seems intuitively reasonable that any method of evaluating probabilistic predictions should give a higher score to the predictor assigning a higher probability to the outcomes that actually occur.

In particular, a reasonable requirement of an evaluation rule is that the score should be related to the magnitude of error in prediction. The statement that the chances of a disease are 10 in 100 means that the disease is 10 times more likely than if the chances were only one in 100. The evaluation score should reflect such a difference in predictions. An evaluation procedure that strictly meets these two requirements is a measure, of predictive accuracy, based on the likelihood assigned to the outcome that occurred [9].*

The formula for the likelihood (L) assigned by an observer to a series of events in which the outcomes are that the event either does or does not occur is†

$$L = \prod_{i=1}^{n} p_i^{x_i} (1 - p_i)^{1-x_i},$$

where n is the number of predictions, p_i the estimated probability that the outcome would occur in the ith case, $x_i = 1$ if the event occurred in the ith case, $x_i = 0$ if the event has not occurred in the ith case, and i a subscript that designates to which of the n cases a value refers (e.g., i = 3 refers to the third case). In words, L is the product of each of the probabilities assigned to the outcome that actually occurred in the n cases.

Since the value of the likelihood L decreases as n, the number of predictions, increases, it is desirable to remove this effect of n by calculating the average value of L per prediction. Calculations are made easier by logarithms, and interpretation is made easier by addition of a constant. The resulting measure provides a coefficient of predictive accuracy designated by the letter Q.

$$Q = \sum_{i=1}^{n} (\log_2 p_i^* + 1)/n,$$

where $p_i^* = p_i$ if the event occurred in the ith case and $p_i^* = 1 - p_i$ if the event did not occur in the ith case. In words, the accuracy coefficient Q is related to the sum of the logarithm of the probabilities assigned to the outcome that actually occurred in each of the n cases. The sum is then divided by n so that the accuracy coefficient is related to the average probability assigned to the outcomes that actually occurred. Properties of logarithmic scoring rules are discussed elsewhere [10–12]. An alternative and equivalent interpretation of the evaluation procedure can be formu-

---

*Statistical usage distinguishes between probabilities and likelihoods. The distinction is a fine one and not essential to the substance of this paper. The term "probability" in this paper refers to the chances (relative frequency) that an outcome *will* occur. The term "likelihood" here refers to the chances assigned by different predictors to a particular outcome that has already occurred.

†In the formulas, the symbol $\sum_{i=1}^{n} p_i$ stands for the summation of all values of p from the first through the nth value. Similarly the symbol $\prod_{i=1}^{n} p_i$ stands for the product of all values of p from the first through the nth value. For example, if $p_1 = 0.10$, $p_2 = 0.20$ and $p_3 = 0.50$, then $\sum_{i=1}^{3} p_i = 0.10 + 0.20 + 0.50 = 0.80$ and $\prod_{i=1}^{3} p_i = 0.10 \times 0.20 \times 0.50 = 0.01$.

lated in terms of the payoff obtainable by "betting" on the occurrence of an outcome in proportion to its estimated probability. Under this interpretation, skill in prediction is provocatively measured by the "earnings" of a physician accumulated over a series of predictions [13].

There is a relation between the accuracy coefficient and the error rate. Intervals can be constructed in which all predictions falling within an interval are assigned the same value. For example, all predictions in the range of 0.10 to 0.20, corresponding to the notion of "rather unlikely," might be assigned a value such as 0.15. When this limited degree of precision is appropriate to the purposes of an evaluation, such "rounding" makes the elicitation and interpretation of probabilities easier. Often, this modification will not substantially affect the rank order of different predictors. The accuracy coefficient corresponds to using an infinite number of intervals; the error rate can be shown to correspond to the case in which the number of intervals is two.

The accuracy coefficient (Q) has the interpretation that a value of 1 equals perfect prediction, a value of 0 indicates no predictive skill, and a value less than 0 implies the existence of negative predictive skill—the predictions made are worse, on the average, than continued noncommittal assertion that any outcome is as likely to occur as not.

Predictive Skill of Physicians

The predictive skill of physicians was examined at the weekly clinical conferences in immunology at Stanford University Medical Center. At these conferences, cases involving the management of rheumatic conditions, including rheumatoid arthritis, neck pain, Wegener's granulomatosis and systemic lupus erythematosus, are presented for evaluation to a group of faculty, fellows, residents and students. After a case presentation, physicians and students were asked to write their views of the probability that a particular clinical outcome would occur. Examples of outcomes in different cases included the probability that a patient would have an increased creatinine clearance of at least 25 ml per minute within two months after the institution of azathioprine therapy, that biopsy of the nasal mucosa would show a granulomatous process, or that serum complement (C3) would be less than 50 mg per deciliter. In each case the outcomes were selected to represent the essential uncertainties underlying a patient's management. The actual outcome was not known at the time of prediction but was learned within several weeks as a patient's course evolved.

A second method of evaluation involved presentation to physicians and students of detailed case histories of patients with systemic lupus erythematosus in which the outcomes were known but not to the participants. Physicians and students submitted predictions on the outcomes that were similar to those used in the weekly conference. Predictive skill was examined by comparison of predictions with the actual patient outcomes by means of the accuracy coefficient.

Results of 11 physicians and students, based on their prediction of outcomes in the cases with systemic lupus erythematosus, are shown in Table 1. In general, predictive skill was closely related to level of training. Faculty scored higher than fellows, who scored higher than residents, who in turn outscored students.

To gain a better feeling for the degree of skill represented by these scores, the scores of these real physicians were compared to the results achieved by a variety of hypothetical physicians, each of whom followed a different strategy. The first hypothetical physician, physician A, followed a strategy of optimistically asserting that unfavorable outcomes occur only 10 per cent of the time, regardless of the type of outcome or of the patient. The second, physician B, was even more optimistic and consistently asserted that unfavorable events occur only 0.1 per cent of the time. The third hypothetical physician (C) was one with extensive knowledge of the literature but with little patient experience. It might be assumed that from his reading he will know the frequency of occurrence of a given event such as death within 12 months among patients with a given disease but still may not be able to modify this average frequency to the unique characteristics of a given patient. The frequency of clinical outcomes such as death, deterioration of renal function, exacerbation of arthritis or onset or exacerbation of serositis was supplied to physician C on the basis of review of the past experience of patients at Stanford.

Table 1. Predictive Accuracy of Faculty, Fellows, Residents and Students*

Physician	Status	Accuracy Coefficient	Error Rate (%)
1	Faculty	0.323	23
2	Faculty	0.291	24
3	Resident	0.288	24
4	Faculty	0.279	24
5	Fellow	0.260	24
6	Faculty	0.221	25
7	Fellow	0.197	24
8	Fellow	0.181	25
9	Resident	0.096	27
10	Student	0.093	26
11	Student	0.039	27

*Each score is based on 100 predictions. Accuracy measured by the accuracy coefficient is compared with accuracy as measured by the error-rate method. The error-rate method failed to detect meaningful differences between faculty & students.

Table 2. Predictive Performance of Five Hypothetical Physicians*

Predictor	Status	Accuracy Coefficient	Error Rate (%)
Specialists	Hypothetical physician D	0.359	23
Fellows	Hypothetical physician E	0.328	23
Physician 1	Faculty	0.323	25
Physician 8	Fellow	0.181	25
Average outcome probability	Hypothetical physician C	0.117	29
Physician 11	Student	0.039	27
"Optimist" (.10)	Hypothetical physician A	−0.071	29
"Optimist" (.001)	Hypothetical physician B	−1.891	29

*When overall average outcome probabilities taken from the literature were used for prediction without regard to patient characteristics, results were poor in comparison with those of the better clinicians. Predictions made without regard to either outcome or patient characteristics did worse when assessed by the accuracy coefficient but were identical according to the error-rate approach. Average probabilities of faculty & fellows were more accurate than the predictions of any single physician. Scores of actual physicians 1, 8 & 11, taken from Table 1, are included for comparison.

Table 2 shows the performance of these three hypothetical physicians in making the same predictions about actual patient outcomes as were made by the physicians in Table 1. The results obtained with the accuracy coefficient are shown in the first column, and those obtained with the error rate in the second. The literature-based physician (C), using only average outcome probabilities, did not do as well as most physicians but did perform better than two students and one resident. The hypothetical "optimistic" physicians (A and B) did far worse than any of the real physicians or students. The error-rate method of evaluation, however, failed to indicate any difference in results between the three different prediction strategies.

It was anticipated that physicians' opinions about the probability of given outcomes would vary. The extent of variation discovered, however, was surprising. Figure 1 shows the judgment of physicians about the probability that the fluorescent-antinuclear-antibody (FANA) titer would be greater than 1:80 in a particular 38-year-old woman with a two-month

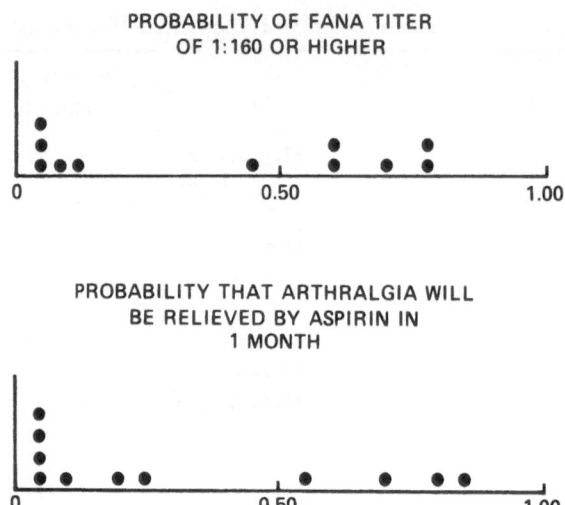

Figure 1. Probabilities of two clinical outcomes as viewed by 11 rheumatologists.

history of arthralgia and a white-cell count of 3400 per cubic millimeter. On several other questions the probabilities asserted by the physicians were apparently uniformly distributed between 0 and 1. The conference discussions concerning these patients failed to indicate such diversity of opinion. When physicians were later individually questioned, however, the diversity of opinion suggested by the probabilities proved real. A tentative conclusion to be drawn from this experience without pretense of rigor is that in discussing the management of a particular case, asking for a numerical statement of the chance of occurrence of an outcome may offer a sensitive and explicit way of demonstrating differences in clinical judgment between physicians.

Two further hypothetical prediction strategies were examined. These were pooled predictions, one representing the calculated average probability of the faculty physicians (hypothetical physician D), whereas the second physician E) used the average probability offered by the fellows. Both the grouped opinion of the faculty and the grouped opinion of the fellows performed better than any single physician (Table 2). This result probably represents the averaging out of extreme opinions. It is an attractive result in appearing to corroborate the benefits obtainable by multiple consultations. However, the composite opinion was only an arithmetic average and was not formed on the basis of an actual group consultation. Experiments examining the benefits of different types and numbers of consultations or the impact of an individual consultant would be informative.

Psychologic studies of probability formation by Kahneman and Tversky [14,15] indicate that individuals tend to use a variety of shortcuts in forming probabilistic assessments. One such shortcut is known as "anchoring." In a medical context, anchoring refers to a physician using the "average" probability of a group as his first estimate of the probability of an event in an individual case. Subsequently, the physician will use additional information concerning the patient to individualize his estimate about this anchor point. Clearly, inaccuracy in prediction can be due either to use of an incorrect anchor point or to failure to individualize appropriately. Skill in these two aspects of prediction is acquired differently. A correct anchor-point probability may be obtained either through knowledge of the literature or by extensive clinical experience. Ability to individualize assessments to the unique characteristics of the patient is primarily a function of experience. Effective individualization of predictions requires consideration of the social and psychologic, as well as biologic, characteristics of a patient. It is possible to partition the accuracy coefficient into a component due to estimation of the average probability and a component due to individualization about this average. Thus, it is possible to study each of these components of predictive skill. These calculations were performed on the predictions of the same physicians whose results are included in Table 1. Table 3 shows a portion of the results. Although the low overall accuracy coefficients obtained by physicians 9 and 10 were almost exactly the same, physician 9 badly esti-

Table 3. Accuracy Coefficients of Four Physicians (Actual Frequency of Events = 0.29)*

Physician	1	2	9	10
Anchor-point probability Source:	0.27	0.14	0.11	0.34
Use of anchor point	0.130	0.023	−0.043	0.122
Individualization around anchor point	0.193	0.268	0.139	−0.029
Accuracy coefficient	0.323	0.291	0.096	0.093

*The accuracy coefficient can be partitioned to show the portion of the score attributable to a physician's mean predicted probability or anchor point & the portion attributable to individualization of estimates around the anchor point. The process is demonstrated with use of the scores of 4 physicians taken from Table 1.

The score due to use of a physician's anchor point represents the accuracy coefficient obtainable by use of his anchor point as the probability of an event in all cases. A constant prediction of 0.34, the anchor point of physician 10, would have resulted in an accuracy coefficient of .122. The additional contribution to predictive accuracy made by a physician's individualization of predictions around his anchor point can be calculated directly but is more easily obtained by subtraction from the overall accuracy coefficient. Physicians 9 & 10 differ widely in their ability to individualize their predictions. The score due to individualization will be >0 when the physician's predictions are more accurate on the average than use of a single mean probability and will be <0 when predictions are less accurate than use of a single well chosen mean value.

mated the overall average of 0.29, and this misestimation accounted for much of his poor score. The difficulty of physician 10, on the other hand, appears to lie in poor individualization.

Applications in Technology Assessment

To make predictions about an individual case, a physician must combine information based on his knowledge of disease with information derived from the clinical examination, the results of laboratory tests and diagnostic procedures and, often, the opinion of consultants. Sources of information that by themselves may not permit accurate prediction may be helpful in improving the predictive accuracy of physicians. Conversely, a source of information that is of unquestionable value by itself may make a negligible contribution over and above the accuracy provided by information that is less costly and less hazardous to obtain.

As an example, one source of information that has recently become available is the computer "consultation." As indicated above, such innovations deserve careful evaluation not only because of their expense but also because of the gravity of the decisions in which such technologies participate. To make this evaluation, two questions were asked. The first was, How accurate in themselves are the assessments made by commonly used methods of computer prediction, and the second, What is the marginal contribution [16] of these assessments to a physician who uses them simply as an additional consultative resource to aid his own predictive thinking?

Twenty years ago, Paul Meehl addressed methodologic issues concerning the first question in his monograph, *Clinical versus Statistical Prediction* [17], and included in his discussion the results of 20 studies, 19 of which found no difference in the predictive skill of clinicians versus that of statistical algorithms. Although computers now make possible approaches of far greater complexity than those considered by Meehl, most studies continue to find the performance of these approaches "comparable" with that of the clinician [18]. The methods of comparison in these studies have invariably been methods based on the error-rate concept.

Adoption of a method based on the likelihood concept as used in the accuracy coefficient provides a sensitive means of comparing the accuracy of predictions made by physicians or by computer. To make this comparison, the predictions of three statistical algorithms were obtained concerning the chances that in each of 50 patients with systemic lupus erythematosus development of the nephrotic syndrome would occur. The computer predictions were based on just a few important immunologic and hematologic determinations (hematocrit, creatinine, initial degree of proteinuria, complement C3, anti-DNA antibody titers and Westergren erythrocyte sedimentation rate). Clinicians were provided case histories summarizing the data in the patient's chart. The three statistical methods for estimating probabilities were Bayes's rule using the assumption that variables are conditionally independent of each other [19], Bayes's rule using a method that takes the correlations between variables into account [20] and a "nearest-neighbor" method that estimates a patient's probability of outcome by finding the frequency of that outcome in the 40 most similar patient histories stored in the computer [3,5].

Results are shown in Table 4. No computer algorithm performed as well as the better clinicians, but these statistical approaches did outperform one student and one resident.

It should not be surprising that the simple but representative computer algorithms used do not outperform the physician, who can use many variables, can take correlation of symptoms into account and can adjust his conceptual structure to the context of a particular patient's illness. In one patient, the severity of a symptom may be the key to prognosis, in another, the rate of change in severity of the symptom, and in a third, the correlation of this symptom to another symptom may be most important. Physicians have and employ a conceptual flexibility that has been lacking in the statistical approaches that have been used in computer diagnosis.

The second question was addressed by comparison of the accuracy of physicians' predictions made before and after obtaining a computer "consultation." Preliminary results revealed that all physicians revised their predictions toward those of the computer. Although movement toward the computer does not necessarily imply an increase in predictive accuracy, all physicians who initially scored lower than the computer improved their accuracy when provided with the computer's assessment.

Table 4. Predictive Accuracy of Physician Versus Computer*

Predictor	Accuracy Coefficient	Error Rate (%)
Physician 1	0.281	20
Physician 2	0.266	24
Physician 3	0.258	23
Physician 4	0.247	24
Physician 5	0.226	24
Physician 6	0.207	24
Nearest neighbor	0.168	24
Bayes theorem (including correlations)	0.145	24
Bayes theorem (independence assumption)	0.121	26
Physician 7	0.084	24
Physician 8	0.046	26

*The skill of 8 physicians was compared with that of 3 representative statistical methods in assessing the probability of nephrotic syndrome in patients with systemic lupus erythematosus. Computer algorithms outperformed a minority of physicians. It is not yet known whether the predictive accuracy of the subspecialty physicians who participated in this study can be regarded as representative of the accuracy of physicians in other settings.

The method of computer prediction was the "nearest-neighbor" approach that gave the results shown in Table 4. The availability of a measure of predictive accuracy makes it possible to evaluate potential improvements in methods of making predictions. In a series of trials, revision of the procedures for selecting groups of similar patients resulted in a "nearest-neighbor" method that scored almost as well as the best physicians tested.

The same approach used to measure the marginal value of computer consultation can be applied to assess the marginal informational value of the results of diagnostic tests such as assays of serum digitalis levels [21]. In this example physicians' predictive accuracy in identifying digitalis toxicity with information gained from consideration of a patient's history, physical examination, electrocardiogram and serum electrolytes could be compared with their predictive accuracy achieved after they learned the results of the serum digitalis level. In focusing on the value of a test to the physician, this method allows the theoretical value of a test to be distinguished from the value of the test as used in practice. A physician who abandons his clinical judgment in deference to the results of a new test that he understands only imperfectly may suffer a decrease in predictive accuracy.

Such considerations suggest that the method may find additional applications in measuring the ability of a physician to use information provided by specialized diagnostic tests. Measurement of a physician's skill in using a given body of information to make predictions is a potentially valuable method of physician assessment that needs to be explored. The advantage of the approach is that it offers a graded means of evaluating the physician's use of information in contrast to the right-or-wrong measure of factual knowledge used in more traditional evaluation methods.

Acknowledgment. Supported by a grant (HS 01875) from the National Center for Health Services Research to the American Rheumatism Association Medical Information System.

I am indebted to Dr. James Fries for advice and support, to Drs. Halsted Holman, Matthew Liang and Robert Pantell for a critical review of the manuscript, to Andrew Dannenberg and Jean Porter for several of the computer analyses described, to Alison Harlow and Catherine Williams for technical assistance and to the members of the Division of Immunology and the San Francisco Bay Area rheumatologists who participated in these studies.

References

[1] Jennett B, Teasdale G, Braakman R, et al: Predicting outcome in individual patients after severe head injury. *Lancet* 1:1031–1034, 1976

[2] Jennett B, Plum F: Data banks for standardized assessments of coma. *N Engl J Med* 295:624, 1976

[3] Rosati RA, McNeer JF, Starmer CF, et al: A new information system for medical practice. *Arch Intern Med* 135:1017–1024, 1975

[4] Stoupel E: Forecasting in Cardiology. New York, Wiley, 1976

[5] Fries JF, Holman HR: Systemic Lupus Erythematosus: A clinical analysis. Philadelphia, Saunders, 1975

[6] Siegler M: Pascal's wager and the hanging of crepe. *N Engl J Med* 293:853–857, 1975

[7] Brier GW: Verification of forecasts expressed in terms of probability. *Monthly Weather Rev* 78:1–3, 1950

[8] Epstein ES, Murphy AH: A note on the attributes of probabilistic predictions and the probability score. *J Appl Meteorol* 4:297–299, 1965

[9] Shapiro AR, Porta J: Comparison of prediction algorithms: physician versus computer. Presented at Ninth International Biometric Conference, Boston, August 23, 1976

[10] Good IJ: Rational decisions. *J R Statist Soc* [B] 14:107–114, 1952

[11] McCarthy J: Measures of the value of information. *Proc Natl Acad Sci USA* 42:654–655, 1956

[12] Winkler RL: The quantification of judgment: some methodological suggestions. *J Am Statist Assoc* 62:1105–1120, 1967

[13] Shapiro AR: A tool for evaluating clinical predictions. Presented at the annual meeting of the Robert Wood Johnson Clinical Scholar Program, Asheville, North Carolina, April, 1976

[14] Kahneman D, Tversky A: On the psychology of prediction. *Psychol Rev* 80:237–251, 1973

[15] Tversky A: Assessing uncertainty. *J R Statist Soc* [B] 36:148–159, 1974

[16] Neuhauser D: Cost-effective clinical decision-making: implications for the delivery of health services, Costs, Risks, and Benefits of Surgery. Edited by JP Bunker, BA Barnes, F Mosteller. New York, Oxford University Press, 1977

[17] Meehl PE: Clinical versus Statistical Prediction: A theoretical analysis and a review of the evidence. Minneapolis, University of Minnesota Press, 1954

[18] Ross P: Computers in medical diagnosis. *CRC Crit Rev Radiol Sci* 3:197–243, 1972

[19] Warner HR, Toronto AF, Veasey LG, et al: A mathematical approach to medical diagnosis: application to congenital heart disease. *JAMA* 177:177–183, 1961

[20] Bahadur RR: A representation of the joint distribution of responses to n dichotomous items, Studies in Item Analysis and Prediction. Edited by H Solomon. Stanford, Stanford University Press, 1961, pp 158–168

[21] Ingelfinger JA, Goldman P: The serum digitalis concentration—does it diagnose digitalis toxicity? *N Engl J Med* 294:867–870, 1976

25

Constructing an EXPERT Knowledge Base for Thyroid Consultation Using Generalized Artificial Intelligence Techniques

Casimir A. Kulikowski and Jack H. Ostroff

A generalized scheme for building consultation systems based on techniques of artificial intelligence (A.I.) was used to construct a sequence of thyroid consultation models. This scheme, called EXPERT [30], provided a language in which the decision making elements and rules of the clinical expert were defined, compiled, and tested against a data base of cases. In the present paper we report on the incremental process of refining the original model through repeated cycles of empirical testing, re-definition, and re-testing. This process was facilitated by the development of programs that interfaced the EXPERT system with the independent thyroid data base, and analyzed performance, thus enabling a rapid assessment of the effect of changes in the decision making rules.

1. Introduction

The problem of thyroid diagnosis provided one of the earliest and most fertile applications of formal decision making and computer aided methods in medicine [5,10,14,15]. The approaches used in the past have included statistical, logical, pattern recognition, and clustering techniques for building consultation models. Several comparisons of these methods have also been carried out. All these techniques draw on the accumulated experiential information from clinical data bases, and have performed well at their development sites.

Some of the major problems limiting the wider applicability and potential benefit of such systems have been: inter-institutional differences in definitions of measurement and diagnostic terms, patient population profiles, and characteristics; faulty assumptions in the application of the

formal models; and the difficulty of easily or rapidly modifying an existing computer program to adapt to a different clinical environment. In
addition, the mathematical processing of probabilities or heuristic uncertainty measures may defy explanation in clinical terms, regardless of
performance.

The alternative approach of encoding in a program the subjective reasoning sequence followed by an expert clinician avoids this latter problem, and serves as the basis for a large number of algorithmic decision
programs. One of the difficulties encountered with this method is that the
same decision may be reached equally well by many different reasoning
paths, and it becomes too rigid and idiosyncratic to settle on a single,
though possibly very effective path. If we are to capture in computer programs some of the flexibility, expressive reasoning power, and explanatory ability of the human expert, new and more powerful techniques for
expressing expert knowledge in the computer will be needed. Many
researchers have been experimenting with more sophisticated variants of
traditional reasoning techniques and data structure implementations for
the past few years [16,26,27,31,32]. In addition, there has been a con-
certed effort by researchers in artificial intelligence to introduce and test
their ideas to help in the design of generalized medical consultation
schemes. This has been part of broader research into expert systems,
which has sought to develop a better understanding of expert human
problem solving methods through computer-based experimentation and
testing. The psychological basis of clinical investigations with cognition
has also been investigated [6].

2. Artificial Intelligence and Knowledge Engineering

In its early days, artificial intelligence concentrated on general problem
solving techniques [11], and resulted in many studies of the strategies
used in stylized problems and game playing with deterministic rules.
Since the early 1970's there has emerged a trend of research into problem
solving in real-life domains, particularly chemistry, medicine, genetics,
and geology. This research attempts to develop an understanding of the
representations of knowledge, to explore the justification and validation
of expert reasoning for situations where the environment is often incom-
pletely defined, partially structured, and presents itself with a considerable degree of uncertainty. Building knowledge bases for such situations
is referred to as knowledge engineering [8]. One such problem solving
situation is that of expert clinical decision making.

Several artificial intelligence methods have been developed over the
past decade for representing the structure of knowledge needed to characterize clinical decisions. They include causal networks [28], production
rules [23], prototypical templates or frames [17], and hierarchical net-

works [18]. Several consultation systems have been built based on the knowledge of clinical experts, or groups of experts, and later tested with substantial numbers of clinical cases. The first major systems were in the areas of ophthalmology—CASNET [29], infectious diseases—MYCIN [21], and internal medicine—INTERNIST [19] and PIP [24]. More recently, consultation systems in other medical areas have been developed which combine representational and inferential ideas from several of the earlier systems [2,3,4,7,20]. There have also been efforts to develop generalized schemes for assisting the expert in rapidly encoding his reasoning concepts and decision rules, developed from two of the early AI systems. These are the EXPERT [30] and EMYCIN [25] schemes that evolved from the CASNET and MYCIN programs, respectively, and the AGE system [12].

3. The EXPERT Representation For Consultation Systems

The EXPERT [30] scheme provides a language, EL for describing a consultation model for a given medical domain. A model can be written in this language using any available text editor. The file containing the model is then compiled by a special program XP, which checks for syntactic inconsistencies, and produces a compact representation for use with the EXPERT consultation program and a data base structure for case information storage. This consultation program prompts for individual case data and produces an interpretational analysis using a simple but effective set of decision strategies. A data base search and analysis program is available to carry out empirical performance analysis for various categories of patients that the investigator can specify. As the result of individual case or group analysis, suggestions for changes in the model can be developed and the EXPERT model modified accordingly. The special facilities that were developed to interface with the existing large thyroid data base, along with the programs for systematizing and easing the updating of the model, are described in section 4. In the remainder of this section we give a brief outline of the EXPERT representation and consultation scheme.

3.1. EXPERT Representational Language (EL)

The goal of this higher-level language is to facilitate the representation of general consultative knowledge bases. The components of the language include a set of conceptual primitives from which reasoning rules can be built.

Conceptual primitives can be: diagnostic hypotheses; pathophysiological state hypotheses, which can be structured according to a taxonomic hierarchy and/or causal network; and prognostic or treatment hypotheses, which can also be related hierarchically or according to sequences of management. The findings or observations are specified according to a typology for data acquisition. Binary (yes/no), multiple exclusive choice, checklist, and numerical value questions are allowed. Mnemonics are used together with an extended English language representation to characterize each hypothesis or finding. The full text is used for display and interaction with users, while the mnemonics provide a shorthand for the specification of reasoning rules.

The normative, or rule-based part of EL includes three kinds of rules. Finding-to-finding (FF) rules express logical constraints among the findings. During a consultation session, a three-valued logical assignment can apply to a finding: it is either true, false, or unknown for the patient visit under consideration. Thus, FF rules express strictly deterministic logical relations. Uncertainty in the results of a finding can be expressed by creating an explicit set of questions about factors affecting confidence in the results, rather than using a generalized uncertainty heuristic.

The major inferences for consultation are performed by finding-to-hypothesis (FH) rules. Various logical combinations of findings serve as the antecedent of a rule. A hypothesis, weighted by a confidence factor, is the consequent part. The confidence values are chosen to separate belief in a hypothesis (on a 0 to 1 scale) from disbelief in the hypothesis (on a 0 to -1 scale). This has been found to be more effective in expressing expert knowledge than a probability measure that links belief and disbelief in a fixed manner [22,28].

To represent patterns of inference and treatment recommendations we also require hypothesis-to-hypothesis (HH) rules. These are designed to allow specification of a context in the form of any pattern of findings and hypotheses which can trigger a set of other rules. The rules within the context can combine findings and hypotheses on the antecedent side, and allow the assignment of a truth value to a single hypothesis as a consequent. The hypotheses in the antecedents can be allowed to take on values in any desired interval. A $(-1:0)$ interval would specify negation of the hypothesis at any level of disbelief, while an interval of $(0.6:1)$ would specify confirmation with a heuristic weight of 0.6 or greater.

3.2. EXPERT Consultation Program

The reasoning of the EXPERT program is invoked in the following sequence:

(a) Initialization: The program user is asked to enter a set of initial findings, or else the knowledge base (model) can be designed to ask for preliminary items, such as the chief complaints or a referring complaint of the patient.

(b) Evaluation of FF Rules: The current set of findings is used to propagate information in a purely deterministic logical manner to other findings. For instance, knowing the sex of the patient will set to FALSE all findings that are associated exclusively with the opposite sex, or knowing that a patient is asymptomatic will rule out asking for the individual, subsumed symptoms.

(c) Hypothesis Generation by the Evaluation of FH Rules: The system evaluates all the FH rules that are triggered by either the initial findings, the response to questions asked by the consultation program, or the ones whose truth values have been established through the FF rules. Since the findings have no confidence factors associated with them, and once true or false for a given patient they remain so, it is simple to partition the entire group of FH rules from the knowledge base into those that are applicable to the patient (by having the required truth values of all findings of the left-hand side combination satisfied), those that are not applicable (by having at least one false member in the F-side), and those that remain of undetermined applicability. The status of applicable or not applicable is a permanent one for a given patient consultation session since the truth values of the findings remain constant (excluding corrections of mistakes, which result in a re-evaluation of the relevant rules).

(d) Hypothesis Generation by Evaluation of HH Rules: Once all FH rules have been evaluated, the consultation system proceeds to generate weights for hypotheses that can be inferred through the HH rules. Only HH rules found in tables which have their IF part evaluated as true are considered. All such HH rules must be reevaluated sequentially after new results of findings are received. The premises and consequents of HH rules may include hypotheses and associated intervals of confidence. Unlike findings which remain true or false, these intervals can change not only directly, but from other rules (both FH and HH) which affect the confidence measure of a hypothesis. HH rules are evaluated in the order of their appearance in the model. There is no backwards chaining, because the order of evaluation is known in advance. Because of this explicit ordering, no assumptions of independence of hypotheses or limits on self-referencing rules are required.

The above procedures result in the assignment of confidence measures which can be directly determined from the rules of evidence (FH) and hypothesis (HH) weight propagation. When more than one rule is applicable, the maximum absolute value of confidence is used. Another procedure is invoked that is helpful both in question selection and as a simple heuristic to adjust weights slightly. Each hypothesis which has some positive evidence, in the form of a satisfied rule or a partially satisfied rule (with unknown truth value) is marked. The count of such rules which apply to each hypothesis is kept. This corresponds approximately to the number of positive indications of the hypothesis.

(e) Hypothesis Generation by Propagation of Taxonomic-Causal Weights: Another mechanism for generating weights for hypotheses is to compute forward and inverse weights propagated through the taxonomic and causal links connecting hypotheses in the descriptive component of the knowledge base. The computation of these weights is based on a generalized version of the CASNET method [28].

Forward weights are propagated from predecessor to successor and inverse weights are propagated from successor to predecessor. A taxonomy contains implied relationships between hypotheses that can be treated similarly to causal connections. The procedures used to generate weights are similar, but not identical, to those used in CASNET [28].

(f) Overall Ranking of Hypotheses: A final weight is derived from the rule-based and taxonomic-causal net weight. It is taken as the maximum absolute value from all the indicated directions (with the appropriate sign). A bonus may be awarded to the final weights. The bonus is given on the basis of the percentage and number of rules (derived directly from findings) that are covered by any single hypothesis. The largest bonus is given to the hypothesis which can potentially cover the most rules. The bonus effect is slight and has its greatest effect when some results of findings have not yet been received and fewer high confidence rules are satisfied. It is most useful when the model contains many rules between single findings and hypotheses, and few or no rules with combined findings in their left-hand side. The bonus can be adjusted by the model designer to have the effect of a scoring function, or, if desired, it can be removed.

At the present stage of development of EXPERT, the overall weight of a hypothesis is computed from the various partial weights described above. Based on this overall weight, a ranking of the various hypotheses can be obtained. Different problem solving strategies can be formulated based on the weights and the characteristics of the hypotheses. These are currently under investigation.

3.3. Program Implementation

The system is written in interactive FORTRAN and occupies about 70K on a DEC-20 computer. FORTRAN provides an increased capability to produce relatively efficient production models and has also enabled it to be implemented on a mini-computer (PDP-11). The program XP compiles a model for use by the EXPERT consultation program, and indicates any errors found in the model.

There are two particularly interesting capabilities of the compiler. First, it can automatically review and convert saved cases to a new format. This allows the modification and updating of a model without loss of time and effort in reentering old cases. A list of added or deleted findings and hypotheses is determined by comparing the mnemonics of the two models.

Secondly, the consistency checking module considers the effect of changes to a model. It then runs through all the stored cases and indicates major modifications in the conclusions that have arisen.

4. Tools for Model Refinement

In AI expert consultation systems, refinement of a diagnostic model often proceeds incrementally, with changes chosen by examination of the model's performance on individual cases. Where overall performance on a large data base of cases is to be optimized, it is common to recreate the entire knowledge base using information from a statistical analysis of the whole data base. What we have done is to make use of empirical information from overall performance on the data base to guide incremental improvements to the model.

In each cycle or iteration, we base our changes largely on two major aspects of the model's performance. First is a comparison of computer and presumptive diagnoses for the entire data base. This gives us some overall idea of the sensitivity and specificity of the model as a whole, and can suggest the need for revision of whole sections of the model or even of the underlying strategy. Further, by selecting subgroups of misdiagnosed cases for closer examination, we can potentially identify smaller sections of the model (groups of rules) which may be responsible for the misdiagnoses.

Second, a profile of the performance of each rule can help determine its role in assigning final computer diagnoses, and whether these are correct or not. These two aspects are highly interrelated, and our justification for basing refinements of the model on this information is basically an extension of the reasoning involved in making incremental changes based on single case performance.

If a case with a given presumptive diagnosis is misdiagnosed by the model, that portion of the knowledge base responsible for assigning the computer diagnosis may be considered suspect. If that portion is a single rule, as is usually the case in EXPERT, it is possible that either the condition for triggering that rule is too weak (the rule is firing when it shouldn't) or the rule's action is too strong (it is overriding the action of some other rule which should in fact produce the final computer diagnosis). On the other hand, it is also possible that the portion of the model responsible for assigning the correct diagnosis may have too strong a triggering condition or too weak an action.

If a specific rule is responsible for such a misdiagnosis in a large number of cases, one can feel more confident that this is the former situation and that changing that rule will have a positive effect on the model's overall performance. If, however, a rule causes only a small percentage of misdiagnoses and is also responsible for many correct diagnoses, the latter situation is more likely, although it is possible the data available to the model may by insufficient for a correct diagnosis. This can sometimes be determined by examining the profile of computer diagnoses for all cases with the relevant presumptive diagnosis.

To improve a model as suggested in response to its overall performance requires that data be available for analyzing the deficits in this performance. We have devoted much effort to developing a group of programs which accomplish this by interfacing our data base of thyroid cases with the EXPERT system. These programs facilitate the gathering and display of a variety of data of value in the model refinement process. They are ultimately intended to be of general utility, requiring minimal effort to allow any EXPERT model to be tested and improved based on its performance on any accessible data base.

While the EXPERT system does have the ability to store and retrieve cases, the storage format is dependent on the model, specifically on the taxonomy and the findings. In many cases, a model is developed only after a suitable data base for testing that model is built up. This is the case with our thyroid data base. Although it would be possible to convert an entire data base into the EXPERT format, in most cases the required storage space would be prohibitive. Also, such an approach would affect the generality and utility of this method, since use with an active data base, as in a medical practice, would effectively require maintaining the data base in duplicate, once the initial conversion was made. Because of this, it was considered preferable to develop the ability to sequentially convert the data base cases into a format suitable for direct 'diagnosis' by EXPERT, with extraction and compilation of only selected information to be displayed for analysis of the model.

At the heart of our interfacing efforts are a group of arrays, used to guide the translation of each data base item into the correct slot and format in an array identical to EXPERT's internal table of findings. In fact, this array is passed to one of EXPERT's routines, exactly as if the data had been entered directly to EXPERT. Since this has now proven itself to be a feasible method of interfacing a large data base to an EXPERT consultation model, the format of the arrays will be made somewhat more general. Currently, the routines which process the arrays and perform the conversion have provisions for conversion of numerical items, yes-no questions and multiple choice lists, inter-type conversions, and the handling of a few special cases. Findings may also have default values or be set according to the value of other findings.

After a case is converted, control is passed to the EXPERT system, which then interprets it in terms of the model and calculates final weights for the hypotheses. These weights are stored, along with the presumptive diagnosis, for each case. This file of final weights is then used to produce a cross listing of computer diagnoses (CDX) versus presumptive diagnosis (PDX):

CDX-PDX	EU	HYPO	HYPER
EU	2316	151	39
HPEP	0	0	0
HPOP	1	23	0
HPES	338	24	283
HPOS	65	267	0

[EU = euthyroid; HYPO = hypothyroid; HYPER = hyperthyroid; HPEP, HPOP = suspected hyperthyroid, hypothyroid; HPES, HPOS = probable hyperthyroid, hypothyroid]

At this point, we have several programs available for displaying data to help choose and support the next set of changes to the model. The most important of these programs enable analysis of individual subgroups of cases, based on computer and presumptive diagnosis. The cross listing mentioned above is used for selection of subgroups of interest. For example, the above table indicates that there are 338 euthyroid cases with a computer diagnosis of suspected hyperthyroidism at this stage in our refinement of the model.

By again interfacing to EXPERT, the activity of each rule can be determined within this group. This data is displayed as a table of the number of times each rule was triggered and the number of times it 'hit' (was responsible for assigning the final computer diagnosis). This enables the identification of rules responsible for the misdiagnosis of a large number of cases, as discussed above. Below is part of the table for those euthyroid cases called hyperthyroid suspects by the model:

Rule	Triggered	Hit	Sets	Weight
98	6	1	HPES	0.90
99	16	14	HPES	0.85
100	84	15	HPES	0.60
101	9	9	HPES	0.95
102	52	39	HPES	0.80
103	45	18	HPES	0.65
104	21	0	HPES	0.50
105	43	0	HPES	0.35

Here we see that Rule 102 caused 39 cases of misdiagnosis out of the 52 cases which satisfied its antecedent. In fact, Rules 98 to 105 as a group were responsible for 96 euthyroid cases being called suspected hyperthyroid by the model. These particular rules are based on the measurement of uptake of radioactive iodine by the thyroid.

It is often useful to be able to have some idea of the effect of changing a rule. If you are considering lowering the weight assigned by that rule, you would like to know if there are any hypotheses with weights high enough that they may override the newly assigned weight. Another of our programs produces a listing of the three top weighted hypotheses for each case in a subgroup. The display is grouped by presumptive diagnosis and sorted by the top hypotheses, i.e. computer diagnosis, within each group.

In the case of the uptake rules, we felt that a high uptake value was actually legitimate grounds for diagnosing a case as probable hyperthyroid (HPEP,) but only if the case was already considered a suspect on clinical grounds. On this basis, Rules 98 to 105 were changed accordingly. After recompiling the model, its performance was as shown:

CDX-PDX	EU	HYPO	HYPER
EU	2448	162	77
HPEP	13	0	172
HPOP	1	22	0
HPES	192	0	73
HPOS	66	281	0

Note that the effects of the change were not limited to the 338 cases, or even to the 96. Correct diagnosis of euthyroid cases increased from 85% to 90%, and of hypothyroid cases from 62% to 66%. Although correct diagnosis of hyperthyroidism dropped from 91% to 79%, many of these cases are now considered probable rather than simply suspect, an improvement in accuracy of diagnosis.

Changes in performance for a model depend on the purpose for which the model is being developed. For example, whether a computer diagnosis of a "suspected" disease state is considered acceptable may not be the same for a model to be used for initial screening purposes as for a model intended to be used for consultation by a generalist clinician.

5. Conclusions

The results of this experiment show that it is easy to convert a large, existing clinical data base of cases into the EXPERT format, which can then provide the basis for model refinement. The expert's original model is tested against cases with known diagnoses, and rules of reasoning that lead to incorrect or partially correct conclusions can be identified and modified. The tools for model analysis and updating that we have developed provide the basis for a methodology of successive empirical testing and model improvement.

Acknowledgments. The authors wish to thank Dr. R. A. Nordyke, of the Straub Clinic, Honolulu, for making available the thyroid data base, and Dr. Sholom Weiss, Mr. Kevin Kern and Mr. Peter Politakis for advice in using the EXPERT system. This work was supported in part by Grant RR-643 of the Biotechnology Resources Program, DRR, NIH.

References

[1] Aikins, J., "Prototypes and Production Rules: An Approach to Knowledge Representation for Hypothesis Formation," Proceedings 6th IJAI, Tokyo; pp. 1–3, 1979.

[2] Blum, R.L. and Wiederhold, G., "Inferring Knowledge from Clinical Data Banks Utilizing Techniques from Artificial Intelligence," Proceedings 3rd Annual Symposium on Computer Applic. in Med. Care, Washington, pp. 303–307, 1979.

[3] Catanzarite, V.A., and Greenburg, A.G., "Neurologist: A Computer Program for Diagnosis in Neurology", Proceedings 3rd Annual Symposium on Computers in Health Care, Washington, pp. 64–71, 1979.

[4] Chandrasekharan, B., Gomez, F., Mittal, S., and Smith, J., "An Approach to Medical Diagnosis Based on Conceptual Schemes," Proceedings 6th IJAI, Tokyo, pp. 134–142, 1979.

[5] Crooks, J., Murray, L.P.C., and Wayne, E.J., "Statistical Methods Applied to the Clinical Diagnosis of Thyrotoxicosis", *Quart. J. of Med.*, v. 110, pp. 211–234, 1959.

[6] Elstein, A.S., Shulman, L.S., and Sprafka, S.A., Medical Problem Solving: An Analysis of Clinical Reasoning, Cambridge, Ma., Harvard Univ. Press, 1978.

[7] Fagan, L.M., Kunz, J.C., Feigenbaum, E.A. and Osborn, J.J., "Representation of Dynamic Clinical Knowledge: Measurement Interpretation in the Intensive Care Unit" Proceedings 6th IJCAI, Tokyo, pp. 260–262, 1979.

[8] Feigenbaum, E.A., "The Art of Artificial Intelligence Themes and Case Studies of Knowledge Engineering," Proceedings National Computer Conference, AFIPS Press, p. 221, 1978.

[9] Kulikowski, C.A., "Pattern Recognition Approach to Medical Diagnosis," IEEE-Transactions on SSC V. 6, pp. 85–89, 1970.

[10] Lusted, L. Introduction to Medical Decision-Making, C. Thomas, Springfield, Ill., 1968.

[11] Newell, A. and Simon, H.A. Human Problem Solving, Prentice Hall, Englewood Cliffs, N.J., 1972.

[12] Nii, H.P. and N. Aiello, "AGE (Attempt to Generalize): A Knowledge-based Program for Building Knowledge Based Programs," Proceedings of the 6th IJCAI pp. 645–655, Tokyo, 1979.

[13] Nilsson, N., Principles of Artificial Intelligence, Tioga Publishing Co., Palo Alto, Ca. 1980.

[14] Nordyke, R., Kulikowski, C.A., and Kulikowski, C.W., "A Comparison of Methods for the Automated Diagnosis of Thyroid Dysfunction", *Computers in Biomedical Research*, v. 4, pp. 374–389, 1971.

[15] Overall, J.E., and Williams, C.M., "Conditional Probability Program for the

Diagnosis of Thyroid Function", *J. Amer. Med. Assoc.,* v. 183, pp. 307–313, 1963.

[16] Patrick, E.A., et. al., "Review of Pattern Recognition in Medical Diagnosis and Consulting Relative to a New System Model," IEEE-Trans. SMC v. 4, pp. 1–16, 1974.

[17] Pauker, S.G., G.A. Gorry, J.P. Kassirer and W.B. Schwartz, "Towards the Simulation of Clinical Cognition: Taking a Present Illness by Computer," *Amer. J. Med.* V. 60, pp. 981–996. 1976.

[18] Pople, H., J. Myers and R. Miller, "DIALOG: A Model of Diagnostic Logic for Internal Medicine," Proceedings 4th IJCAI, Tbilisi; p. 841, 1975.

[19] Pople, H., "The Formation of Composite Hypotheses in Diagnostic Problem Solving," Proceedings 5th IJCAI, Boston; pp. 1030–1037, 1977.

[20] Reggia, J.A., "A Production Rule System for Neurological Localization," Proceedings 2nd Annual Symposium on Computer Application in Medical Care, Washington, pp. 254–260, 1978.

[21] Shortliffe, E.H., et al., "An Artificial Intelligence Program to Advise Physicians Regarding Antimicrobial Therapy," *Computers in Biomedical Research* v. 6; pp. 544–560, 1973.

[22] Shortliffe, E.H. and B. Buchanan, "A Model of Inexact Reasoning in Medicine," *Mathematical Biosciences,* v. 23, pp. 351–379, 1975.

[23] Shortliffe, E.H., Computer-based Medical Consultations: MYCIN, Elsevier, 1976.

[24] Szolovitz, P. and S.G. Pauker, "Categorical and Probabilistic Reasoning in Medical Diagnosis," *Artificial Intelligence,* v. 11, pp. 115–144, 1978.

[25] vanMelle, W., "A Domain Independent Production-Rule System for Consultation Programs," Proc. of the 6th IJCAI; pp. 923–925, Tokyo, 1979.

[26] Warner, H.R., "Knowledge Sectors for Logical Processing of Patient Data in the HELP System," Proceedings 2nd Annual Symposium on Computer Application in Medical Care, Washington, pp. 401–404, 1978.

[27] Wechsler, H., "A Fuzzy Approach to Medical Diagnosis," *Internat. J. Biomed. Comp.,* v. 7, pp. 191–203, 1976.

[28] Weiss, S., C. Kulikowski, S. Amarel, and A. Safir, "A Model-Based Method for Computer-Aided Medical Decision-Making," *Artificial Intelligence* v. 11, pp. 145–172, 1978.

[29] Weiss, S., C.A. Kulikowski and A. Safir, "Glaucoma Consultation by Computer," *Comp. Biol. Med.* v. 8, pp. 24–40, 1978.

[30] Weiss, S. and C. Kulikowski, "EXPERT: A System for Developing Consultation Models," Proceedings of the 6th IJCAI, Tokyo; pp. 942–950, 1979.

[31] Woodbury, M. and Clive, J., "Data Based Definitions of Disease: Suggestions for a solution of the formal diagnostic problem," Proceedings 13th HICSS, Honolulu, pp. 590–600, 1980.

[32] Wortman, P.M. "Medical Diagnosis: An Information Processing Approach," *Computer Biomed. Res.* V. 5, pp. 315–328, 1972.

26

Towards an Intelligent Textbook of Neurology

James A. Reggia, Thaddeus P. Pula,
Thomas R. Price, and Barry T. Perricone

We define an intelligent textbook of medicine to be a computer system that:
(1) provides for storage and selective retrieval of synthesized clinical knowledge for reference purposes; and (2) supports the application by computer
of its knowledge to patient information to assist physicians with decision
making. This paper describes an experimental system called KMS (a
Knowledge Management System) for creating and using intelligent medical
textbooks. KMS is domain-independent, supports multiple inference methods and representation languages, and is designed for direct use by physicians during the knowledge acquisition process. It is presented here in the
context of the development of an Intelligent Textbook of Neurology. We
suggest that KMS has the potential to overcome some of the problems that
have inhibited the use of knowledge-based systems by physicians in the
past.

1. Introduction

Background

During the last several years a significant research effort has been devoted
to developing computer systems for processing clinical knowledge. Such
systems are referred to as "knowledge-based" programs because they are
built upon a database of medical knowledge. The knowledge they contain
is usually "synthesized" in the sense that it has been prepared by one or
more medical experts from multiple existing information sources (textbooks, the literature, personal experience, etc.).

The predominant approach to creating knowledge-based systems has been focused primarily on computer use of knowledge to assist in medical decision making [Reggia, 1979]. While those developing such computer-assisted *medical decision*-making (CMD) systems have paid some attention to providing explanatory and educational abilities, most of the effort in this work has concentrated on the decision-making process itself. In particular, the problems of knowledge representation (How do we represent medical knowledge in terms of abstract data structures that are processable by machine?) and automatic inference generation (How do we use these abstract data structures in the context of a particular patient to make decisions and produce useful information?) have been the central issues. Creating knowledge-based CMD systems has been difficult and time consuming, and providing the supporting software requires techniques that challenge the current state of the art in computer science.

An alternative but less frequently pursued approach to developing medical knowledge-based systems has centered on simply storing and retrieving free text information. The motivation behind this work is the desire to provide physicians with a computerized medical reference system. For example, the National Library of Medicine has developed a prototype Hepatitis Knowledge Base along these lines for practicing physicians [Bernstein et al, 1978]. Ten experts on this topic organized the relevant information that went into the knowledge base in a modular, hierachical fashion that facilitated its selective retrieval.

It is our belief that both of these approaches to developing medical knowledge-based systems are useful, and furthermore, that their combination in a single system would be very profitable. In other words, it would be of value to have a system composed of synthesized knowledge bases that (1) could be read by physicians for its informational content just like a textbook, and (2) could be applied by computer to information about a specific patient to generate a differential diagnosis, estimate a prognosis, stage a disease process, recommend laboratory tests or potential therapy, and so forth. In this report we will use the term *intelligent textbook of medicine* to describe such an interactive knowledge-based computer system. Physicians would interact with intelligent textbooks as either authors (after a training period), users (after a brief explanation), or both.

Creating an intelligent textbook of medicine involves formidable problems. For example, information in each of its knowledge bases would need to be readable by physicians, yet at the same time be sufficiently formalized so that it could be processed by machine to assist with decision making. In addition, an efficient method for acquiring and modifying knowledge bases would be necessary in view of the large amounts of clinical knowledge that would be involved. In this paper we present an approach to developing an intelligent textbook of medicine that addresses these and other requirements.

An Intelligent Textbook of Neurology

We are currently investigating the feasability of developing a prototype Intelligent Textbook of Neurology. We feel that Neurology is a particularly appropriate domain in which to work since it is relatively circumscribed compared to medicine in general, yet it is still sufficiently rich to involve a broad range of issues. Our research is emphasizing the translation of existing neurological knowledge in textbooks and the literature as well as new information derived from clinical databanks into a form suitable for computer processing. We are especially interested in improving the physician-computer interaction in the knowledge acquisition process.

In this paper we report on our progress to date in this endeavor. We begin by describing a system called KMS that supports the development and use of intelligent textbooks of medicine as we defined them above. Two simple examples of using KMS to create knowledge bases that contain neurological problem-solving information are then presented. Following these examples further details of the architecture of KMS are described. Finally, we conclude with a discussion of why our approach to developing an intelligent textbook at least has the potential to overcome some of the problems that have inhibited the use of knowledge-based systems by physicians in the past.

2. A Knowledge Management System

An Overview of KMS

KMS (a *K*nowledge *M*anagement *S*ystem) is an experimental computer system for developing intelligent textbooks that is currently being studied at the University of Maryland. In other words, KMS is not an intelligent textbook itself, but rather a tool for use by physicians or others interested in authoring and using intelligent textbooks. The term "management" is employed in the name 'KMS' to emphasize that the use of medical knowledge (i.e., the selection of an appropriate inference method, the specific inferences to make) as well as its representation is of concern.

The ultimate goal is for KMS to provide all of the necessary software (but not knowledge) for an intelligent textbook in a prepackaged form. A physician-author would use KMS as a "workbench," selecting the appropriate components for a particular task without having to do any programming. All that would be required from the author is a knowledge base written in one of the formal, non-procedural KMS languages (see below). Ideally, KMS would therefore be usable by previously computer-inexperienced physicians who wished to author part of an intelligent textbook after only a brief training period.

To achieve this goal, KMS has been designed with the following three key features:

(1) KMS is domain independent.

Most of the research into CMD technology during the last several years has concentrated on developing "dedicated" computer programs that are concerned with a fixed range of medical problems (e.g., infectious diseases [Shortliffe, 1976], internal medicine [Pople et al, 1975], abdominal pain [deDombal, 1975]). In contrast, KMS is not designed for any specific medical domain but is meant to be as general and widely applicable as possible. In this sense KMS bears a resemblance to some other research efforts currently in progress. For example, the rule-oriented software originally developed for MYCIN (called EMYCIN for "Essential MYCIN") has been used to build CMD systems in several different areas of medicine [van Melle, 1979]. A second domain-independent rule-based system is EXPERT, a generalized descendent of the CASNET formalism [Weiss and Kulikowski, 1979]. It is currently being used to develop consultation programs in rheumatology, ophthalmology, and endocrinology. Finally, another rule-based system for building consultation programs is AGE [Nii and Aiello, 1979]. AGE has been used to implement a consultation system dealing with pulmonary function test interpretation. While KMS is related to these and other domain-independent systems, the next two key features distinguish it from them.

(2) KMS provides multiple inference methods and knowledge representation formats.

Previous CMD systems for assisting with specific medical decision-making tasks have generally been based on a single representation format and inference method. For example, Bayesian systems (e.g., [deDombal, 1975]) have represented the program's knowledge as a collection of probabilities, and then made inferences based on Bayes' Theorem, Production systems (e.g., [Shortliffe, 1976]) have a knowledge base composed of a set of conditional rules, and make decisions by using an antecedent or consequent-driven rule interpreter as an inference method. Each of the domain-independent systems mentioned above (AGE, etc.) is also oriented around a single formalism for managing knowledge. One can find reports in the literature describing why each of these inference methods and representation formats is the most useful one for developing CMD systems.

In contrast, KMS is composed of a collection of subsystems, each of which is based on a different inference method and representation format. This approach to the architecture of KMS reflects the belief that there is no single "best" method for representing and using knowledge. On the contrary, there is a variety of methods that are available, each with certain advantages and disadvantages. The selection of which method to use for a given problem depends on several factors such as the structure of the problem involved and the availability of appropriate problem-solving knowledge.

(3) KMS is for direct use by physicians during the knowledge acquisition process.

KMS is based on the belief that the best way to develop medical knowledge-based systems is by permitting a physician to transfer *directly* his or her knowledge to the computer. Thus, KMS supports formal representation languages for creating knowledge bases whose primitive elements are the attributes, values, and associations of a particular domain problem. These languages are meant to be comprehensible to physicians who have had little or no prior computing experience.

Again, this feature of KMS contrasts with the more traditional viewpoint [Amarel, 1977; Feigenbaum, 1977] that a computer scientist should serve as an intermediary between the physician-author and the computer, helping the medical expert to express his or her problem-solving knowledge (see Figure 1a). The KMS viewpoint is that the computer scientist (or "knowledge engineer") should serve in a different capacity dealing predominantly with epistemological issues (Figure 1b). Specifically, such an individual would create and modify the KMS subsystems, educate physician-authors and users about the system, and be available for consultation as specific knowledge bases are developed.

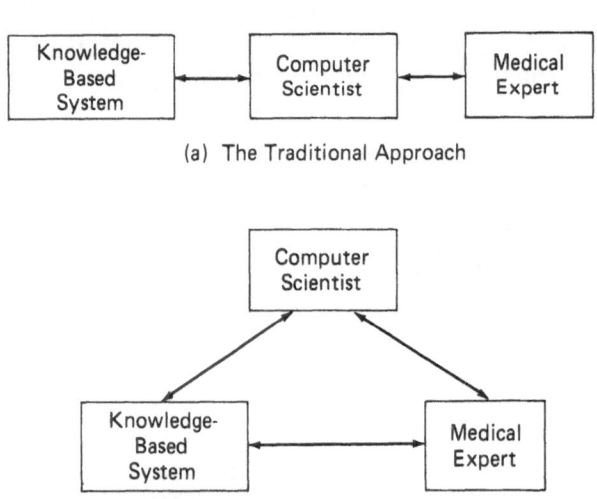

(a) The Traditional Approach

(b) The KMS Approach

Figure 1. Alternate views of the computer scientist's role.

(1) Construct a problem-oriented inference network
(2) Select an inference method
(3) Encode the knowledge base
(4) Evaluate and certify the knowledge base

Figure 2. The four-step process for creating a knowledge base.

Creating a Knowledge Base

A physician who wants to create a knowledge base for inclusion as part of an intelligent medical textbook goes through a four step process (Figure 2). First, the physician must conceptually organize the underlying knowledge in terms of a 'problem-oriented inference network,' a data structure that serves as a framework for writing a knowledge base. Second, he or she then selects an inference method (e.g., Bayes' Theorem, deduction) that will be used in conjuction with the knowledge base. Third, the actual encoding of the knowledge base is done using the appropriate KMS language. Finally, the implemented knowledge base is evaluated for accuracy and usefulness prior to its inclusion as part of the intelligent textbook.

3. Examples of Knowledge Base Acquisition

It will be useful at this point to present a few examples of how a physician uses KMS to author knowledge bases. We intend to illustrate the four step process outlined immediately above and to reinforce our emphasis on the translation of existing medical information into formal knowledge bases that are processable by machine. The examples are meant to be simple and thus they do not illustrate many features available in KMS. Other examples involving prognosis following cardiopulmonary resuscitation, thyroid function test interpretation, and lung cancer staging can be found in [Reggia, 1980].

Example 1: Stroke Diagnosis

A stroke is the sudden onset of a neurological deficit (e.g., unilateral paralysis, coma, loss of ability to communicate) due to disruption of the brain's blood supply. In terms of frequency, stroke is one of the top three causes of death and disability in the United States today. Knowledge about stroke would form an important part of any Intelligent Textbook of Neurology.

When faced with a patient who has just experienced a stroke, the physician must make several important decisions. Central to these decisions is the question of which specific disease process has caused the stroke. The possible diagnoses include thrombotic infarction (blockage of a cerebral artery by a thrombus), lacune (mild stroke due to very small vessel disease), embolic infarction (blockage of a cerebral artery by an embolus), intracerebral hematoma (bleeding into the brain following blood vessel rupture), and subarachnoid hemorrhage (bleeding around the outside of the brain due to blood vessel rupture). The clinical classification of a patient into one of these diagnostic categories must be as accurate as possible because the appropriate treatment varies from category to category. The selection of an inappropriate treatment can lead to disastrous results (e.g., anticoagulation for a patient with an intracerebral hematoma).

Approximately two years ago an article about stroke diagnosis appeared in the neurological literature [Mohr et al, 1978]. This particular article (hereafter referred to as the 'reference article'), was based on the results of a several year study involving about 700 patients. It reported the frequency of each stroke category (prior probabilities) and the frequency of relevant signs and symptoms in patients with each type of stroke (conditional probabilities). To create a knowledge base from this article for inclusion in our Intelligent Textbook of Neurology, we would go through the following four step process.

The first step is to conceptually construct a *problem-oriented inference network* or, more simply, an *attribute hierarchy,* that will provide the underlying organization for the knowledge base. The term 'problem-oriented' indicates that each such network is centered around a specific domain problem. The *attributes* (nodes) in the network are those features about a patient that are of importance in making decisions. For example, these attributes might include the patient's AGE, SEX, TEMPERATURE, or MOTOR EXAMINATION RESULTS. Conceptually associated with each attribute is a set of possible *values* which that attribute can have. The attribute SEX, for example, can take on the possible values MALE or FEMALE, while the attribute MOTOR EXAMINATION RESULTS might take on the values LEFT HEMIPARESIS, RIGHT HEMIPARESIS, or NORMAL. The links between the attributes in the network represent the relation "depends on" (or conversely, "determines") indicating that the value of one attribute (drawn in a superior position) is determined by the values of others (drawn in inferior positions). Thus, the lowest level of an attribute hierarchy consists of *input attributes* whose values are determined by a user; other nodes in the network represent *inferred attributes* whose values are determined by the system. The most general form of a problem-oriented inference network is a directed, acyclic graph as shown in Figure 3. Attributes A1 through A5 are input attributes while the rest are inferred attributes.

For our stroke example there is only one inferred attribute called TYPE OF STROKE (see Figure 4). Its value is determined by the values of twelve input attributes found to correlate statistically with stroke diag-

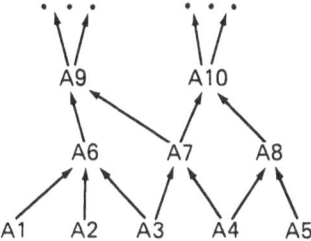

Figure 3. Abstract representation of a problem-oriented inference network.

nosis in the reference article. These include whether or not there is a HEADACHE AT ONSET OF STROKE, the TIME COURSE OF STROKE EVOLUTION, the RESULTS OF LUMBAR PUNCTURE, and several other factors. Although it is not shown in Figure 4, conceptually associated with each of these attributes is a set of possible values. For example, the possible values of TYPE OF STROKE include THROMBOTIC INFARCTION, LACUNE, etc.; those of TIME COURSE OF STROKE EVOLUTION include SUDDEN ONSET, STEPWISE OR STUTTERING ONSET, etc. The precise definitions of these and other attributes and their values are given in the reference article. By expressing knowledge in terms of a problem-oriented inference network like this, one creates a non-procedural framework for knowledge that is both understandable to computer-inexperienced clinicians and yet amenable to formal representation in a knowledge base that can be processed by computer.

The second step in developing the knowledge base is to select an appropriate inference method. In our stroke example, the values of TYPE OF STROKE are mutually exclusive. In addition, only probabilistic inferences are involved, with the relevant probabilities being provided in the reference article. Thus, a reasonable inference method to use would be Bayes' Theorem [Duda and Hart, 1973]. It is necessary to falsely assume that the values of input attributes are independent in this case, but experience has shown that Bayes' Theorem will generally give good results as long as we eliminate those attributes that are obviously depen-

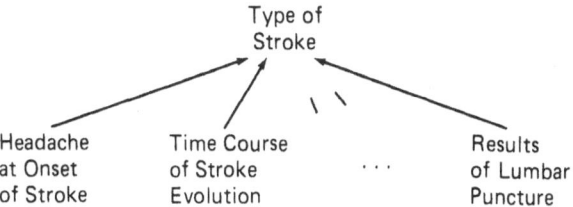

Figure 4. A simple, one level attribute hierarchy for stroke diagnosis.

dent [Salamon et al, 1976]. For example, although PAST HISTORY OF HYPERTENSION correlates with the TYPE OF STROKE, it is left out of the knowledge base while the presumably more relevant HYPERTENSION ON EXAMINATION would be included. We will also assume that the values of TYPE OF STROKE are the only types of strokes although this is not strictly true.

The third step in developing the knowledge base is to do the actual encoding. For our purposes we will view knowledge bases as consisting of just two parts (there are others): a knowledge schema and an associations section. In its simplest form, the schema consists of a list of the names of the attributes in the attribute hierarchy along with the names of their possible values. For example, the input attribute TIME COURSE OF STROKE EVOLUTION in Figure 4 and its possible values would be declared in the schema as

TIME COURSE OF STROKE EVOLUTION:
    SUDDEN ONSET
    / STEPWISE OR STUTTERING ONSET
    / SMOOTH OR GRADUAL ONSET
    / FLUCTUATIONS BETWEEN NORMAL AND ABNORMAL AT
ONSET.

This is illustrated in Figure 5 which is the entire encoded knowledge base.

The association section of a knowledge base, which may overlap in part with the schema, provides information that associates the possible values of different attributes to one another. For our Bayesian stroke diagnosis knowledge base, this simply consists of the appropriate prior and conditional probabilities given along with the values of the inferred attribute TYPE OF STROKE (see Figure 5). The prior probabilities are listed in parentheses immediately following the inferred values; the conditional probabilities of input attribute values are then listed in the same order as they appear in the schema. While not illustrated here, it is possible to have more than one inferred attribute, each of which depends on a subset of the input attributes.

The fourth and final step in developing the knowledge base is its evaluation. The first part of the evaluation is done by KMS: given a knowledge base, KMS will examine it for errors much as a compiler examines a high-level language program (e.g., a FORTRAN program). If errors are found by KMS then explanatory error messages are generated and the knowledge base is rejected. Both syntactic errors and a limited number of semantic errors (e.g., the prior probabilities of the values of an inferred attribute do not sum to 1.0) can be detected. If no errors are found, the knowledge base is accepted and transformed into an internal interpretable form.

The rest of the knowledge base evaluation involves testing the accuracy of the knowledge base's recommendations, predictions, etc. in practice. (This is a difficult task whose description is beyond the scope of this article.) A certification history is then kept as part of each knowledge base. This history records who originally wrote the knowledge base, what modifications have susbsequently been made to it, and what testing it has undergone.

Once a knowledge base has been organized, written and tested, it can be cataloged as part of an intelligent textbook. Thereafter an individual desiring to use that knowledge base can sign onto KMS, instruct the system to 'activate' the knowledge base, and then use the available commands to accomplish various tasks. For example, a user can direct KMS to OBTAIN ⟨attribute⟩. This command tells KMS to use the currently active knowledge base to determine the value of the specified attribute. Another command is DISPLAY ⟨option⟩. This command allows the user to view part or all of a knowledge base or the internal representation of a problem that is being solved. A third example is NEXT ⟨option⟩. This command tells KMS that the user wants to discuss a new problem (NEXT CASE) or that the system should switch to a new knowledge base (NEXT KB).

An example of an interaction between a user and KMS is shown in Figure 6. The dialog begins at the point where the user has just activated our example stroke knowledge base. Items typed by the user are underlined for clarity. This conversation was made possible by supplying KMS with only the knowledge base shown in Figure 5; it was unnecessary for the knowledge base author to do any computer programming in the conventional sense.

Example 2: Periodic Paralysis

Periodic paralysis is a rare disorder characterized by episodes of diffuse, transient muscle weakness without impairment of consciousness, sensation or coordination. Because of its rarity, many physicians do not remember all of the detailed information needed to reach a diagnosis and to establish treatment. Thus, knowledge about periodic paralysis is a potentially useful component of an Intelligent Textbook of Neurology. In view of space restrictions, we will limit ourselves in this article to discussing the knowledge needed for the diagnosis and treatment of those types of periodic paralysis classified as primary (i.e., hypokalemic, paramyotonic, normokalemic, and thyrotoxic periodic paralysis) and even simplify this knowledge somewhat.

To create a knowledge base of the problem of periodic paralysis, we must go through the same four step process that was outlined in the previous example (Figure 2). First, using reference materials [Adams and Victor, 1977; Buruna and Schipperheyn, 1979] we would construct an

*INPUT ATTRIBUTES

HEADACHE AT ONSET OF STROKE:
    PRESENT / ABSENT.
TIME COURSE OF STROKE EVOLUTION:
    SUDDEN ONSET /
    STEPWISE OR STUTTERING ONSET /
    SMOOTH OR GRADUAL ONSET /
    FLUCTUATIONS BETWEEN NORMAL AND ABNORMAL AT ONSET.
OCCURRENCE OF SEIZURES SINCE ONSET OF STROKE: PRESENT / ABSENT.
OCCURRENCE OF VOMITING SINCE ONSET OF STROKE: PRESENT / ABSENT.
ATHEROSCLEROSIS FACTORS PRESENT:
    -BRUIT -PREVIOUS MI -ANGINA
    -CLAUDICATION OR -ABSENT PULSE
    / NONE OF THESE FACTORS.
TIA HISTORY: YES / NO.
PRESENCE OF DIABETES MELLITUS:
    PRESENT / ABSENT.
PRESENCE OF ATRIAL FIBRILLATION:
    PRESENT / ABSENT.
PRESENCE OF VALVULAR HEART DISEASE:
    PRESENT / ABSENT.
SIGNIFICANT HYPERTENSION ON EXAMINATION:
    PRESENT / ABSENT.
LEVEL OF CONCIOUSNESS ON EXAMINATION:
    COMATOSE / NON-COMATOSE.
RESULTS OF LUMBAR PUNCTURE: BLOODY CSF / NON-BLOODY CSF.

*INFERRED ATTRIBUTE

TYPE OF STROKE(HEADACHE AT ONSET OF STROKE;

        .

        .

        .

RESULTS OF LUMBAR PUNCTURE):

LARGE ARTERY THROMBOTIC INFARCTION(0.34):
        0.12 0.88;
        0.40 0.34 0.13 0.13;
        0.004 0.996;
        0.10 0.90;
        0.56 0.44;
        0.50 0.50;
        0.26 0.74;
        0.08 0.92;
        0.08 0.92;
        0.59 0.41;
        0.04 0.96;
        0.00 1.00 /

Figure 5. A simple KMS.BAYES knowledge base. Lines beginning with '*' are comments.

LACUNE(0.19):
    0.03 0.97;
    0.38 0.32 0.20 0.10;
    0.00 1.00;
    0.05 0.95;
    0.37 0.63;
    0.23 0.77;
    0.29 0.71;
    0.05 0.95;
    0.05 0.95;
    0.77 0.23;
    0.00 1.00;
    0.02 0.98 /
EMBOLIC INFARCTION(0.31):
    0.09 0.91;
    0.79 0.11 0.05 0.05;
    0.04 0.96;
    0.04 0.96;
    0.34 0.66;
    0.11 0.89;
    0.13 0.87;
    0.34 0.66;
    0.25 0.75;
    0.46 0.54;
    0.00 1.00;
    0.01 0.99 /
INTRACEREBRAL HEMATOMA(0.10):
    0.33 0.67;
    0.34 0.03 0.63 0.00;
    0.06 0.94;
    0.51 0.49;
    0.11 0.89;
    0.08 0.92;
    0.15 0.85;
    0.06 0.94;
    0.03 0.97;
    0.91 0.09;
    0.24 0.76;
    0.70 0.30 /
ANEURYSM OR ARTERIONVENOUS MALFORMATION(0.06):
    0.78 0.22;
    0.80 0.03 0.14 0.03;
    0.07 0.93;
    0.47 0.53;
    0.05 0.95;
    0.07 0.93;
    0.02 0.98;
    0.02 0.98;
    0.00 1.00;
    0.34 0.66;
    0.24 0.76;
    0.94 0.06 %

Figure 5. (Continued)

.
.
.

))) ENTER COMMAND:

obtain type of stroke.

))) HEADACHE AT ONSET OF STROKE:
    (1)   PRESENT
    (2)   ABSENT
  = ? *2*

))) TIME COURSE OF STROKE EVOLUTION:
    (1) SUDDEN ONSET
    (2) STEPWISE OR STUTTERING ONSET
    (3) SMOOTH OR GRADUAL ONSET
    (4) FLUCTUATIONS BETWEEN NORMAL AND ABNORMAL AT ONSET
  = ? *1*

.
.

))) RESULTS OF LUMBAR PUNCTURE:
    (1) BLOODY CSF
    (2) NON-BLOODY CSF
  = ? *2*

))) TYPE OF STROKE =
  LARGE ARTERY THROMBOTIC INFARCTION: 0.73
  LACUNE: 0.14
  EMBOLIC INFARCTION: 0.13
  INTRACEREBRAL HEMATOMA: 0.00
  ANEURYSM OR ARTERIOVENOUS MALFORMATION: 0.00

))) ENTER COMMAND:

display type of stroke = lacune.

VALUE: LACUNE
OF: TYPE OF STROKE
PRIOR PROBABILITY: 0.19
CONDITIONAL PROBABILITIES:
  HEADACHE AT ONSET OF STROKE:
    PRESENT: 0.03
    ABSENT: 0.97
  TIME COURSE OF STROKE EVOLUTION:
    SUDDEN ONSET: 0.38

.
.

  RESULTS OF LUMBAR PUNCTURE:
    BLOODY CSF: 0.02
    NON-BLOODY CSF: 0.98

Figure 6. Using the information in the example stroke knowledge base.

```
))) ENTER COMMAND:

next case.

        .
        .
        .
```

Figure 6. (Continued)

appropriate hierarchy (see Figure 7). In this case we want KMS to provide us with one or more possible PARALYSIS TYPEs, a recommendation for the INITIAL MANAGEMENT of the patient, and a suggestion as to what steps might be useful for FUTURE PROPHYLAXIS to prevent further attacks. We would need several key facts about the patient to determine the values of these inferred attributes (SERUM K+, presence or absence of MYOTONIA, etc.).

The next step is to select an appropriate inference method. For our periodic paralysis problem we will elect to use primarily deductive decision making based on conditional rules or productions. Thus, we will encode the knowledge in the reference materials as a collection of rules containing KMS *statements.* An explicit KMS statement is of the form

$$\langle attribute \rangle \; \langle relation \rangle \; \langle value \rangle$$

such as

$$SERUM \; K+ \; = \; HIGH.$$

Statements like this are combined in rules of the form

$$\langle rule \; name \rangle$$
$$if \; \langle antecedent \; statements \rangle$$
$$then \; \langle consequent \; statements \rangle$$

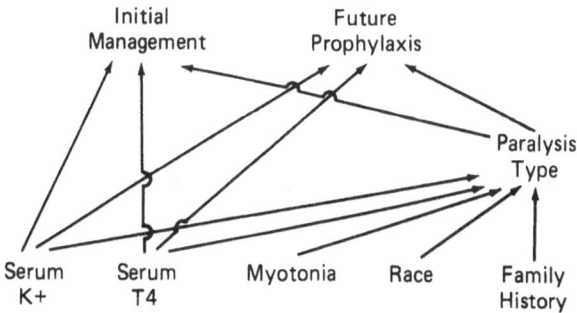

Figure 7. Problem-oriented inference network for primary periodic paralysis.

such as

```
LK2
IF   SERUM K+ = LOW
  &  MYOTONIA = ABSENT
  &  SERUM T4 = HIGH
THEN
      PARALYSIS TYPE = THYROTOXIC (0.8)
  &  PARALYSIS TYPE = HYPOKALEMIC (0.2).
```

This rule is named LK2, and it asserts that *if* a patient with periodic paralysis has a low serum potassium, no myotonia, and a high serum T4, *then* it follows that the diagnosis is probably thyrotoxic periodic paralysis, although hypokalemic periodic paralysis is still a possibility. The numbers at the end of the consequent statements indicate how certain the individual who is writing the rule feels that they follow from the antecedents. These "certainty factors" are derived from either the reference material or personal knowledge.

A complete but simplified and uncertified knowledge base for periodic paralysis is shown in Figure 8. Like the previous example, this knowledge base is divided into a schema (beginning with 'ATTRIBUTES:') and an associations section (beginning with 'RULES:'). The correspondence between these rules and the structure of the attribute hierachy shown in Figure 7 should be studied carefully. The symbol '&' in rules indicates conjunction, while '/' indicates disjunction. The relation '#' signifies 'not equal to' (several other relations exist, such as 'GE' for 'greater than or equal to' for numeric-valued attributes).

A knowledge base consisting of rules like the ones in Figure 8 can be processed by the consequent-driven rule interpreter that is part of KMS. This component of KMS uses a top-down, depth-first strategy to procedurally interpret rules and make inferences. The model of inexact reasoning originally introduced in MYCIN [Shortliffe, 1976] is used to propagate "certainty factors" from a rule's antecedents to its consequents.

We have not illustrated an interactive session involving the periodic paralysis knowledge base because it is very similar to the interaction shown in Figure 6 from the user's point of view. The commands used to control the interaction (e.g., OBTAIN, NEXT, DISPLAY) are basically the same, and a JUSTIFY command provides a simple explanation capability.

4. The Architecture of KMS

Having presented some simple examples of the use of KMS in the knowledge acquisition process, we now turn to a more detailed description of

* PRIMARY PERIODIC PARALYSIS RULE SET
ATTRIBUTES:
FAMILY HISTORY: NEGATIVE / POSITIVE / UNKNOWN;
RACE: ORIENTAL / NON-ORIENTAL;
MYOTONIA: PRESENT / ABSENT;
SERUM K +: LOW/ NORMAL / HIGH;
SERUM T4: LOW OR NORMAL / HIGH / UNKNOWN;
PARALYSIS TYPE:
    HYPOKALEMIC /
    THYROTOXIC /
    PARAMYOTONIC /
    NORMOKALEMIC /
    UNRECOGNIZED;
INITIAL MANAGEMENT:
   REPEAT LAB TESTS TO CONFORM RESULTS
   / COUNTERACT HIGH K + WITH CA GLUCONATE
   / OBTAIN SERUM T4
   / MONITOR CARDIORESPIRATORY STATUS
   / RESOLVE THYROTOXICOSIS
   / KCL 5GM PO;
FUTURE PROPHYLAXIS:
   ACETAZOLAMIDE 250MG PO DAILY
   / MAINTAIN RESOLVED THYROTOXICOSIS
   / UNDETERMINED.

RULES:
LK 1
IF SERUM K + = LOW
  & MYOTONIA = ABSENT
  & SERUM T4 = LOW OR NORMAL
THEN
    PARALYSIS TYPE = HYPOKALEMIC;

LK2
IF SERUM K + = LOW
  & MYOTONIA = ABSENT
  & SERUM T4 = HIGH
THEN
    PARALYSIS TYPE = THYROTOXIC (0.8)
  & PARALYSIS TYPE = HYPOKALEMIC (0.2);

LK3
IF SERUM K + = LOW
  & MYOTONIA = ABSENT
  & SERUM T4 = UNKNOWN
  & RACE = ORIENTAL
THEN
    PARALYSIS TYPE = HYPOKALEMIC (0.5)
  & PARALYSIS TYPE = THYROTOXIC (0.5);

LK4
IF SERUM K + = LOW
  & MYOTONIA = ABSENT

Figure 8. The primary periodic paralysis knowledge base (KMS.PS).

```
    & SERUM T4 = UNKNOWN
    & RACE = NON-ORIENTAL
THEN PARALYSIS TYPE = HYPOKALEMIC (0.9)
    & PARALYSIS TYPE = THYROTOXIC (0.1);


LK5
IF SERUM K+ = LOW
   & MYOTONIA = ABSENT
   & SERUM T4 = UNKNOWN
   & FAMILY HISTORY = POSITIVE
THEN
    PARALYSIS TYPE = HYPOKALEMIC (0.8)
    & PARALYSIS TYPE = THYROTOXIC (0.2);


LK6
IF SERUM K+ = LOW
   & MYOTONIA = ABSENT
   & SERUM T4 = UNKNOWN
   & FAMILY HISTORY = NEGATIVE
THEN
    PARALYSIS TYPE = HYPOKALEMIC (0.3)
    & PARALYSIS TYPE = THYROTOXIC (0.7);


LK7
IF SERUM K+ = LOW
   & MYOTONIA = ABSENT
   & SERUM T4 = UNKNOWN
   & FAMILY HISTORY = UNKNOWN
THEN
    PARALYSIS TYPE = HYPOKALEMIC (0.6)
    & PARALYSIS TYPE = THYROTOXIC (0.4);


LK8
IF SERUM K+ = LOW
   & MYOTONIA = PRESENT
THEN PARALYSIS TYPE = UNRECOGNIZED:


HK1
IF SERUM K+ = HIGH
   & SERUM T4 # HIGH
THEN
    PARALYSIS TYPE = PARAMYOTONIC;


HK2
IF SERUM K+ = HIGH / SERUM K+ = NORMAL
   & SERUM T4 = HIGH
THEN
    PARALYSIS TYPE = UNRECOGNIZED;


NK1
IF SERUM K+ = NORMAL
   & MYOTONIA = ABSENT
THEN
    PARALYSIS TYPE = NORMOKALEMIC (0.7)
    & PARALYSIS TYPE = PARAMYOTONIC (0.3);
```

Figure 8. (Continued)

NK2
IF SERUM K+ = NORMAL
  & MYOTONIA = PRESENT
  & SERUM T4 # HIGH
THEN
    PARALYSIS TYPE = PARAMYOTONIC;

REC1
IF PARALYSIS TYPE = UNRECOGNIZED
THEN INITIAL MANAGEMENT = MONITOR CARDIORESPIRATORY STATUS
  & INITIAL MANAGEMENT = REPEAT LAB TESTS TO CONFIRM RESULTS
  & FUTURE PROPHYLAXIS = UNDETERMINED;

REC2
IF PARALYSIS TYPE = PARAMYOTONIC
  / PARALYSIS TYPE = NORMOKALEMIC
THEN INITIAL MANAGEMENT = MONITOR CARDIORESPIRATORY STATUS
  & FUTURE PROPHYLAXIS = ACETAZOLAMIDE 250MG PO DAILY;

REC3
IF PARALYSIS TYPE = PARAMYOTONIC
  & SERUM K+ = HIGH
THEN INITIAL MANAGEMENT = COUNTERACT HIGH K+ WITH CA GLUCONATE
  & FUTURE PROPHYLAXIS = ACETAZOLAMIDE 250MG PO DAILY;

REC4
IF PARALYSIS TYPE = THYROTOXIC
  & SERUM T4 = HIGH
THEN INITIAL MANAGEMENT = MONITOR CARDIORESPIRATORY STATUS
  & INITIAL MANAGEMENT = KCL 5GM PO
  & INITIAL MANAGEMENT = RESOLVE THYROTOXICOSIS
  & FUTURE PROPHYLAXIS = MAINTAIN RESOLVED THYROTOXICOSIS;

REC5
IF PARALYSIS TYPE = THYROTOXIC
  & SERUM T4 = UNKNOWN
THEN INITIAL MANAGEMENT = MONITOR CARDIORESPIRATORY STATUS
  & INITIAL MANAGEMENT = KCL 5GM PO
  & INITIAL MANAGEMENT = OBTAIN SERUM T4
  & FUTURE PROPHYLAXIS = UNDETERMINED;

REC6
IF PARALYSIS TYPE = HYPOKALEMIC
  & PARALYSIS TYPE # THYROTOXIC
THEN INITIAL MANAGEMENT = MONITOR CARDIORESPIRATORY STATUS
  & INITIAL MANAGEMENT = KCL 5GM PO
  & FUTURE PROPHYLAXIS = ACETAZOLAMIDE 250MG PO DAILY;

REC7
IF PARALYSIS TYPE = HYPOKALEMIC
  & PARALYSIS TYPE = THYROTOXIC
THEN
    INITIAL MANAGEMENT = MONITOR CARDIORESPIRATORY STATUS
  & INITIAL MANAGEMENT = KCL 5GM PO
  & FUTURE PROPHYLAXIS = UNDETERMINED.

Figure 8. (Continued)

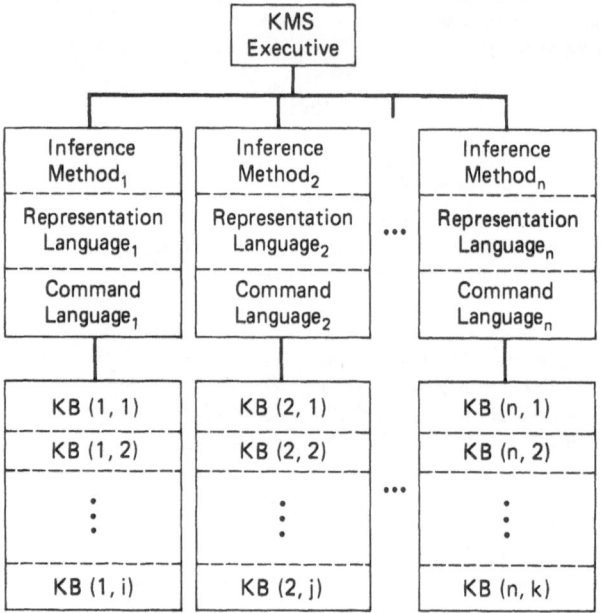

Figure 9. KMS architecture.

the architecture of KMS itself (Figure 9). Basically, KMS consists of a collection of n domain-independent subsystems that are overseen by the KMS executive. Associated with each subsystem is a collection of knowledge bases (KB's).

Structural Details

Each KMS subsystem is organized around a different inference method. All of the programs needed to implement a CMD system using that inference method are available as part of the subsystem. Thus, each KMS subsystem is a collection of software that implements the following components:

Inference System. When called into action this component will apply a stored knowledge base selected by the user to a specific problem. It makes inferences about that particular case that are of interest to the user. For example, if the underlying inference method of a subsystem was Bayes' Theorem, this component would consist of the programs that performed the actual calculations.

Knowledge Base Parser and Interpreter. This component parses a knowledge base written in the appropriate KMS representation language and transforms it into an internal interpretable form. It is responsible for detecting syntactic and a limited number of semantic errors in a knowledge base. It is analogous to a parser in a standard programming language compiler and performs the same useful services.

Knowledge Base Compiler. This component stores a transformed knowledge base in its internal form. Later retrievals of that knowledge base do not require re-examination for errors or re-transformation to internal form and therefore are more efficient. The compiler is intended for use with completed, error-free knowledge bases that are being cataloged as part of an intelligent textbook.

User Interface. This component provides for interaction with a user. It includes the command language which is used to direct the control of the system, display part or all of a knowledge base, etc. In addition, by using information in a knowledge base, this component requests problem-specific information from a user in a simple, natural language-like format.

Together these components implement the interactions illustrated in the examples in the previous section.

Associated with each subsystem is a library of knowledge bases written in the corresponding representation language. These knowledge bases are not part of KMS itself but form the contents of an intelligent textbook. They are provided by the textbook's author(s).

Finally, the KMS executive (KMS.EXEC) is responsible for overseeing the operation of the entire system. It permits an individual to direct the control of KMS without having to worry about the programming details that are involved. For example, the executive will allow one to "browse" through the library of knowledge bases in an intelligent textbook. Once a particular knowledge base is selected, it will then activate that knowledge base and automatically load the appropriate subsystem into the computer to permit use of the knowledge base.

The Complexity of KMS

In considering the development of an intelligent textbook, a reasonable question is whether the complexity of the underlying software would be prohibitive. KMS attempts to answer this question in two ways.

First, significant attention has been given to making KMS as modular as possible. Individuals developing medical knowledge-based systems in the past have stressed the importance of creating modular knowledge bases (intra-knowledge base modularity). KMS has adopted this philosophy and in addition expanded it to include modularity on all levels. Thus, each KMS knowledge base associated with a subsystem as part of an intelligent textbook is an independent unit of medical problem-solving information (inter-knowledge base modularity). Each subsystem is composed of well-defined components (intra-subsystem modularity), and the entire KMS software package is partitioned into independent subsystems (inter-subsystem modularity). This approach to the architecture of KMS has the advantage that only a single problem-oriented knowledge base and subsystem are needed in memory at any time. In addition, it has permitted the incremental implementation of KMS.

The second way in which the complexity question has been handled within the framework of KMS is by making the system relatively easy to learn to use. This involves not only the ideas of user-cordiality [Shneiderman, 1980], but also a major emphasis on inter-subsystem compatibility. Regardless of which subsystem is being used, the underlying conceptual structure of a knowledge base is always built around a problem-oriented inference network as described above. In addition, the information in a knowledge base is always encoded in terms of implicit or explicit KMS statements. The commands required to use a knowledge base are essentially the same in each subsystem. Finally, the user interfaces of KMS subsystems all ask the user for information about a specific problem in the same way. The name of an input attribute whose value is desired by KMS is printed out and the user simply selects from among its possible values in responding. This multiple-choice format is useful for constraining the inferences made by the system [Rieger, 1978], presupposes no special typing skill of the user, and avoids the need for a sophisticated natural language interface.

Current Implementation Status

The first versions of two KMS subsystems have been implemented. One of these is based on Bayes' Theorem (KMS.BAYES) and the second is a consequent-driven production rules sytem (KMS.PS). These were illustrated above. A third subsystem (KMS.HT) is currently being implemented. Knowledge in this latter subsystem is organized into 'frames' and a hypothesize-and-test approach to decision making has been taken. All programming is in non-compiled LISP because of the evolving nature of KMS. The system is available on the University of Maryland Instruction and Research Network (Univac 1100/40). It can be accessed statewide from University campuses or by remote site terminals via telephone.

We have already developed a few functioning prototype knowledge bases such as those demonstrated earlier in this paper. At present we are attempting to demonstrate that physicians can understand the concepts behind KMS and that they can learn to use it to develop knowledge bases after a brief training period. The available KMS subsystems are also currently being used on a regular basis by medical students to do class projects assigned in an elective course on biomedical computing (see [Hybl and Reggia, 1980]). The modular design of KMS has permitted us to use individual subsystems as soon as they are brought on-line without waiting for the completion of the entire system.

Although we have primarily illustrated the use of KMS knowledge bases in support of decision making in this paper, it should be emphasized that KMS is also oriented towards displaying the information contained in knowledge bases. The DISPLAY command can be used to view all or part of any knowledge base (for example, see Figure 6). We anticipate that this will be a powerful reference and educational tool, especially in conjunction with KMS.HT where knowledge is organized in a textbook-like fashion. In addition, the JUSTIFY and TRACE commands (not illustrated here) will be useful along these lines. Finally, we are exploring the incorporation of free text into knowledge bases as a means of supplementing their decision-making information.

5. KMS and the Problems of CMD

In spite of twenty years of research, CMD and other knowledge-based systems have had relatively little effect on the day-to-day practice of medicine. There are several reasons for this and they have been discussed extensively in the literature (e.g., [Croft, 1972; Friedman, 1977; Mitchell, 1970; Startsman, 1972]). Some of the problems that have inhibited the development and use of CMD systems are not directly addressed by KMS: the lack of adequate databases with relevant clinical information, the lack of standardization in medical definitions, etc. However, KMS was designed to alleviate four of the problems that are frequently mentioned.

(1) The physician-computer interaction has not been successfully accomplished.

A distinguishing feature of KMS is that it is designed for direct use by physicians, both as users and as textbook authors. In fact, the idea of direct physician-computer interaction is a central theme of our work. Significant attention is being given to the important 'human factors' that this entails.

(2) The translation of clinical knowledge into a form suitable for computer processing and the implementation of the program to process it are difficult and time consuming tasks.

Each KMS subsystem includes all the software needed to implement a complete CMD system (inference method, user interface, etc.) once it is given an appropriate knowledge base with which to work. Thus no additional programming is necessary. Knowledge acquisition remains a significant undertaking, but it is at least improved by the direct use of formal representation languages by physicians. In addition, KMS expedites knowledge acquisition by providing a collection of languages suitable for different types of knowledge and by detecting certain classes of errors in knowledge bases.

(3) There has been a lack of acceptance of CMD systems by the medical community.

This in part reflects the fact that physicians have not been convinced that CMD systems can be generally useful. Often the emphasis by research workers has been on producing systems that do what the physician does, and not surprisingly this has met with only limited acceptance by physicians. We hope that this problem will be alleviated within the framework of KMS by stressing *support* for decision making rather than by producing computer-generated decisions. Also, since KMS is directly usable by physicians, it gives a physician the freedom to implement a knowledge base that is of specific interest to him or her. In the long run this may prove to be one of KMS's most important features. Finally, by providing a library of knowledge bases, KMS has the potential for accumulating the "critical mass" of information that would be necessary to justify the time required to learn its use. This is not true with conventional knowledge-based systems that address only one problem with a single knowledge base.

(4) Even when successful most CMD systems cannot conveniently be transferred to different installations.

While KMS is theoretically portable to any facility that supports the LISP language, the numerous dialects of LISP make this a less than ideal prospect. Eventually a standardized and portable verion of KMS could be developed. However, the real issue is not the portability of KMS itself, but whether or not knowledge bases (*not* programs) can be made easily transferable from on installation to the next. Since KMS supports machine-independent representation and command languages, any computer facility with an implemented version of KMS would be able to use KMS knowledge bases developed at other sites. While a great deal of work remains to be done, the ideas behind KMS at least have the potential to provide a qualitative improvement in the portability of CMD systems.

6. Conclusion

At present KMS and our concept of an Intelligent Textbook of Neurology are best characterized as experiments-in-progress with questions remaining to be answered about their ultimate utility. For example, will the use of simple inference methods significantly limit the power of the KMS subsystems? It has previously been suggested that "the problem solving power exhibited in an intelligent agent's performance is primarily a consequence of the specialist's knowledge employed by the agent, and only very secondarily related to the generality and power of the inference method employed" [Feigenbaum 1977]. If one accepts this belief then the ability of KMS to manage libraries of knowledge bases gives reason for optimism. Is the direct physician-computer interaction really feasible? The growing diffusion of computer technology throughout society and the increasing computer-sophistication of individuals in medicine makes this a real possibility.

These and other questions about KMS will be examined through the development of an Intelligent Textbook of Neurology. The present plan is to complete KMS.HT and KMS.EXEC, and then to write a small prototype library of neurological knowledge bases. Our emphasis will be on stroke diagnosis and management. The use of KMS by computer-inexperienced individuals (medical students, physicians, etc.) will continue to be evaluated. We will further evolve KMS in the context of the feedback that we receive from those making use of it, and we will explore several additional research questions. Hopefully, KMS or similar systems will ultimately help to make the computer a useful tool for a number of individuals for whom it has previously been relatively inaccessible.

Acknowledgments. The research described in this third report of the NEUREX project is supported by NINCDS through grants 5 KO7 NS 00348 and 1 PO1 NS 16332. Computer time has been provided in part by the Computer Science Center of the University of Maryland. The authors wish to thank Dr. Charles Rieger for his constructive criticism and suggestions.

References

Adams R and Victor M: *Principles of Neurology,* McGraw-Hill, 1977, 953–957.

Amarel S et al: Applications of Artificial Intelligence (Panel), *Proc. Fifth IJCAI,* 1977, 994–1006.

Bernstein L, Seigel E, and Ford W: The Hepatitis Knowledge Base Prototype, *Proceedings of the Second Annual Symposium on Computer Applications in Medical Care,* Washington, D.C., Nov. 1978, 366–367.

Buruma O and Schipperheyn J: Periodic Paralysis, in *Handbook of Clinical Neurology,* Vinken P and Bruyn G (editors), Vol. 41, North-Holland, 1979, 147–174.

Croft D: Is Computerized Diagnosis Possible?, *Comp. Biomed. Res.,* 5, 1972, 351–367.

deDombal F: Computer-Assisted Diagnosis of Abdominal Pain, in *Advances in Medical Computing,* Rose and Mitchell (editors), Churchill-Livingston, 1975, 10–19.

Duda R and Hart P: *Pattern Classification and Scene Analysis,* John Wiley and Sons, 1973.

Feigenbaum E: The Art of Artificial Intelligence—Themes and Case Studies of Knowledge Engineering, *Proc. Fifth IJCAI,* 1977, 1014.

Friedman R and Gustafson D: Computers in Clinical Medicine—A Critical Review, *Comp. Biomed. Res.,* 10, 1977, 199–204.

Hybl A and Reggia J: An Educational Program in Medical Computing for Clinicians and Health Scientists, *Proceedings of the Second National Educational Computing Conference,* Norfolk, Virginia, June 23–25, 1980, 276–280.

Mitchell J: The Automation of Clinical Diagnosis, *Bio-Med. Comp.,* 1, 1970, 157–166.

Mohr J et al: The Harvard Cooperative Stroke Registry—A Prospective Registry, *Neurology,* 28, 1978, 754–762.

Nii H and Aiello N: AGE (Attempt to Generalize)—A Knowledge-Based Program for Building Knowledge-Based Programs, *Proc. Sixth IJCAI,* 1979, 645–655.

Pople H, Myers J and Miller R: A Model of Diagnostic Logic for Internal Medicine, *Proc. Fourth IJCAI,* 1975.

Reggia J: Computer-Assisted Medical Decision Making—Knowledge Bases, *Proc. Third Annual Symposium on Computer Application in Medical Care,* IEEE, Oct. 1979, 28.

Reggia J: A Domain-Independent System for Developing Knowledge Bases, in *Proceedings of the Third Biennial Conference of the Canadian Society for Computational Studies of Intelligence,* University of Victoria, Victoria, B.C., May 14–16, 1980, 289–295.

Rieger C: The Importance of Multiple Choice, TINLAP-II, University of Illinois, July 1978.

Salamon R et al: Bayesian Method Applied to Decision Making in Neurology—Methodological Considerations, *Meth. Inform. Med.,* 15, 1976, 174–179.

Shneiderman B: *Software Psychology: Human Factors in Computer and Information Systems,* Winthrop Publishers, 1980.

Shortliffe E: *Computer-Based Medical Consultations—MYCIN,* American Elsevier, 1976.

Startsman T and Robinson R: The Attitudes of Medical and Paramedical Personnel Toward Computers, *Comp. Biomed. Res.,* 5, 1972, 218–227.

van Melle W: A Domain-Independent Production Rule System for Consultation Programs, *Proc. Sixth IJCAI,* 1979, 923–925.

Weiss S and Kulikowski C: EXPERT—A System for Developing Consultation Models, *Proc. Sixth IJCAI,* 1979, 942–947.

Knowledge Structure Definition for an Expert System in Primary Medical Care

Werner Horn, Walter Buchstaller, and Robert Trappl

Introduction

In the primary medical care sector, an expert consultation system for diagnosis and treatment recommendation has to face a specific situation: in the initial consultation the patient presents a very unspecific collection of subjective findings, the availability of diagnostic procedures is very limited, the physician usually concludes a 'suspected diagnosis,' but he has to decide on correct therapy, including the decisions 'wait and do nothing,' 'send patient to hospital immediately,' and 'send patient for special examination.' The system must take care of follow-up visits, follow the course of disease(s), and supervise the therapeutic effects and side effects, perhaps revising the initial diagnosis in the light of new manifestations and therapeutic outcomes.

The first task in primary care is to detect 'preventable dangerous courses of diseases.' Therefore the possible courses of a disease, together with appropriate medical actions at each stage, and expectations about the future course have to be represented in the system. This has been accomplished for the restricted area of glaucoma (CASNET [8]), but an expert system in primary medical care will often face an unknown or not well know pathophysiology.

Knowledge Structure Definition

The representation of the medical knowledge should serve as a basis for the consultation system performing its task under the requirements men-

Reprinted from the *Proceedings of the Seventh International Joint Conference on Artificial Intelligence*, pp. 850–852, 1981. Used by permission of the International Joint Conference on Artificial Intelligence, Inc.; copies of the Proceedings are available from William Kaufmann, Inc., 95 First Street, Los Altos, CA 94022, U.S.A.

tioned above. Compared to the clinical area, it is hardly possible to include all facets of primary care a priori in an expert system. Decisions will be based e.g. on nosological, etiological, topographical, pathogenetic knowledge. We will discover that some of the necessary facets are not represented after the construction of the knowledge base and after the consultation system has been completed. This fact led us to build the Knowledge Structure Definition System KSDS for easy modification and supplement of knowledge. This idea of developing systems for constructing consultation systems is also stressed by Weiss and Kulikowski [9].

One possibility would be to represent knowledge in rules like MYCIN [7]. However, defining explicit hierarchies for the topography of diseases and for abnormal function of physiological systems is extremely difficult. Moreover, Reggia [6] reported the inadequacy of the rule-based approach for his neurological localization problem.

We designed our knowledge base as a net structure, bearing in mind the design principles of CASNET [8], INTERNIST [4,5], PIP [3], ABEL [2], and CENTAUR [1].

Medical knowledge is represented in a conceptual net. We have defined four different concepts up to now:

- Manifestations
- Diseases
- Diagnostic procedures
- Therapeutic procedures

KSDS represents concepts as nodes connected by relations. Nodes and relations are represented in the same format (compare Figures 1 and 2). Their description in KSDS consists of four parts.

1. The *definition part* (first two lines of Figures 1 and 2) contains

- an unambiguous identification of the node or relation {1},
- the name of the defined entry {2}, and
- if it is a relation: the type of the nodes connected {3,4}.

2. The *descriptive part* declares the attributes of the node or relation ('DESCRIPTORS' in Figures 1 and 2). It is similar to the notion of slots in the frame terminology. The descriptive part for a relation contains values associated with this relation, e.g. strength, preference, consequences,

RELATION {1}	Name {2}			from{3}	to{4}	
TREA	perform therapeutic procedure			D	T	
DESCRIPTORS:						
Name{1}		Ab{2}	Val{3}	Len	Pos	Def{4}
01 necessity of procedure		NEC	NECESS	4	1	1
02 strategy if not effective		NEFF	STNEFF	4	5	1

Figure 1. Definition of a relation.

NODE {1}   Name {2}
  T   Therapeutic procedure

DESCRIPTORS:

Name {1}	Ab{2}	Val{3}	Len	Pos	Def{4}
01 Term of procedure	NAME	C80	80	1	0
02 Classification code	CC	C10	10	81	0
03 Type of procedure	TYPE	T.TYPE	4	91	2
04 Risk	RISK	RISKS	4	95	1
05 Cost	COST	COSTS	4	99	99
06 Prescription	PRES	C80	80	103	0
07 Timelag after becoming effective	LAG	TIMLAG	4	183	99
08 Maximal number of repetitions	MAX	INT	2	187	1

RELATIONS:

Typ Name	Len to-node
01 COV is a possible therapy for disease	4 D
02 CNTR contraindication (incl.allergy)	8 D
03 SIDE side effect (diseases)	8 D
04 MCOV symptomatic therapy for	4 M
05 MCNT contraindication (manifestation)	8 M
06 MSID side effect (manifestation)	8 M

EXPLANATIONS:
Add in the explanation part information about prescription habits and additional information for the patient.

Figure 2. Definition of a node.

expectations using the relation. Thus, Figure 1 shows the relation 'perform therapeutic procedure,' connecting a 'Disease'- with a 'Therapeutic procedure'-node.

Apart from the name and a classification code (ICD, SNOMED), the descriptive part of a node contains items like risk, duration, frequency. Figure 2 shows the complete definition of the node 'Therapeutic procedure.' The information for each of the descriptors consists of the name of the descriptor {1}, its abbreviation {2}, value {3} and default value {4}. In addition, information about the internal structure created by KSDS is displayed (length, position). The value determines a group from a list of allowed primitives (Figure 3: primitives of the group 'types of therapeutic procedures (T.TYPE)'), defined prior to the definition of the concept. The default value will be used, if the expert will not supply a value filling the concept. The usage of primitive-lists allows the checking of plausible values as well as the definition of attributes in natural language.

3. The *relation part* defines all relations starting at this concept. Figure 2 shows the defined relations for the node 'Therapeutic procedure.' Length and to-node are taken from the description and definition part of the relation definition in KSDS.

4. The *explanation part* contains comments on the defined concept

GR	EL	ABBREV	NAME
12		T.TYPE	types of therapeutic procedures
12	1	ADVICE	advice/psycho-sozial therapy
12	2	MEDIC	medication
12	3	PHYSIO	physiotherapy
12	4	RADIO	radiotherapy
12	5	NURSE	nursing care
12	6	FUNCT	substitution of function
12	7	INTENS	intensive care
12	8	SURG	surgical care

Figure 3. List of primitives.

1 Term of procedure: immobilization of affected limb
2 Classification code: 61.2
3 Type of procedure: physiotherapy
4 Risk: minimal
5 Cost: low
6 Prescription: surgical dressing (bandages, casts), usually 2–6 weeks
7 Timelag after becoming effective: immediately
8 Maximal number of repetitions: 0

The element is connected to others by the following relations:
1 ⟨is a possible therapy for disease⟩
    [1] sprain and strain of supportive structures (840–848)
    [2] dislocation of joint (830–839)
    [3] fracture (829)
    [4] cutanous and subcutanous infection (680–86)
    [5] burn (940–949)
    [6] superficial injury (912–917)
2 ⟨contraindication (incl.allergy)⟩
    [1] phlebothrombosis (451)
        1 STR : strong

  .
  .

6 ⟨side effect (manifestation)⟩:
    [1] hypersensitivity reaction to dressing material
        1 EXP : rarely
    [2] pressure sore
        1 EXP : rarely
    [3] disuse atrophy
        1 EXP : rarely

Explanations for the node "immobilization of affected limb":
  The dressing should stay on site until the schedule follow-up contact. In case of pain
and/or swelling an earlier revision is indicated.

Figure 4. Example of a 'therapeutic procedure'-node.

and detailed information for the expert entering new knowledge. It can be used to support a natural language conversation.

Example

The example in Figure 4 shows a 'Therapeutic procedure'-node in the knowledge base.

Concluding Remarks

The advantages of the knowledge structure definition system are:

- There is an explicit and visible definition of the knowledge structure;
- This structure can be modified interactively;
- The internal representation is determined and controlled by the KSDS;
- Modifications in the knowledge structure can easily be performed on already existing knowledge bases. The KSDS will automatically restructure the existing knowledge base;
- The system for the acquisition of knowledge from experts and for consistency checking can use substructures of the net activated for a specific task (e.g. entering information about relations between manifestations and diseases);
- The procedural knowledge of the consultation system will be defined as operations on this knowledge structure, explicitly showing the concepts, descriptors and relations used;
- This will be highly important in defining executors to perform specific tasks during a consultation (e.g. activate diseases indicated by topographical relations as hypotheses, determine possible treatments, check contraindications, etc.). Only one or very few tasks will be active at one moment, corresponding to the standard procedure of a physician in primary care (try to find answers to the following questions: "What is the complaint?", "What does it refer to?", "What should be considered?", "Possible causes?", "What course is expected?", "Do's and don't's?"). Propagation in the net will be restricted by using only very few relations when performing one task. This will (hopefully) prevent combinatorial explosure of search effort;
- One criterion for our consultation system is conceptual adequacy. Presumably, the newly defined executors will perform worse compared to the performance of a physician. Step-by-step modifications of the executors will hopefully increase performance. Since an executor shows the net structure it uses, modification can be done more easily.

References

[1] Aikins J.S., "Prototypes and Production Rules: an Approach to Knowledge Representation for Hypothesis Formation," Proc. 6th IJCAI, Tokio, Japan, 1979.

[2] Patil R.S., "Design of a Program for Expert Diagnosis of Acid Base and Electrolyte Disturbances", MIT/LCS/TM-132, Cambridge, Mass., USA, 1979.

[3] Pauker S.G., Gorry G.A., Kassirer J.P., Schwartz W.B., "Towards the Simulation of Clinical Cognition: Taking the Present Illness by Computer," *Am. J. Med.,* 60, 981–996, 1976.

[4] Pople H.E., Myers J.D., Miller R.A., "DIALOG: a Model of Diagnostic Logic in Internal Medicine," Proc. 4th IJCAI, Tbilisi, USSR, 1975.

[5] Pople H.E., "The Formation of Composite Hypotheses in Diagnostic Problem Solving: an Exercise in Synthetic Reasoning," Proc. 5th IJCAI, Cambridge, Mass., USA, 1977.

[6] Reggia J.A., "A Production Rule System for Neurological Localization," Proc. 2nd Ann. Symp. Comp. Appl. Med. Care, IEEE, 1978.

[7] Shortliffe E.H., "Computer-Based Medical Consultations: MYCIN," Elsevier, New York, 1976.

[8] Weiss S.M., Kulikowski C.A., Safir A., "Glaucoma Consultation by Computer," *Comput. Biol. Med.,* 8, 25–40, 1978.

[9] Weiss S.M., Kulikowski C.A., "EXPERT: A System for Developing Consultation Models," Proc. 6th IJCAI, Tokio, Japan, 1979.

Studying Hypotheses on a Time-Oriented Clinical Database: An Overview of the RX Project

Robert L. Blum and Gio C.M. Wiederhold

The RX computer program examines a time-oriented clinical database and attempts to derive a set of (possibly) causal relationships. First, a Discovery Module uses lagged, nonparametric correlations to generate an ordered list of tentative relationships. Second, a Study Module uses a small knowledge base (KB) of medicine and statistics to create a study design to control for known confounders. The study design is then executed by an on-line statistical package, and the results are automatically incorporated into the KB as a machine-readable record. In determining the confounders of a new hypothesis the Study Module uses previously "learned" causal relationships.

Introduction

One of the most important reasons for accumulating patient data on computers is the possibility of deriving medical knowledge from the stored observations. The long range objectives of the RX Project are 1) to increase the validity of medical knowledge derived from large time-oriented databases containing routine, non-randomized clinical data, 2) to provide knowledgeable assistance to a research investigator in studying medical hypotheses on large databases, and 3) to fully automate the process of hypothesis generation and exploratory confirmation. The objective of this paper is merely to provide a brief overview of the RX project. We strongly urge the interested reader to review references [1] and [2] for annotated examples of the program and for detailed descriptions of its components. Methods for storing and displaying the clinical data are described in [3]. A discussion of the difficulties in drawing inferences from clinical databases appears in [4].

Designed to emulate standard methods of epidemiological research, the RX computer program is a prototype system for automating the discovery, confirmation, and incorporation of knowledge from large clinical databases. While the program is a research prototype, parts of it have been operational since 1979, and have been demonstrated at national conferences.

A medical researcher enters a causal hypothesis of interest into the RX Study Module, for example, "does aspirin decrease blood hemoglobin?" The Study Module uses a small on-line knowledge base (KB) of medicine and statistics to produce a study design of this hypothesis. In doing this, the Study Module also uses pre-computed information on the amount of data on each variable stored in the database. This study design is then executed by a statistical package using the appropriate data from the database. The results of this study are then automatically encoded into the on-line medical knowledge base in a machine-readable form. The KB also contains knowledge that was entered directly into it by clinicians. Both kinds of knowledge are accessible to the Study Module while it is designing a study.

Now, instead of obtaining the initial hypothesis from a medical researcher, it is easy to imagine deriving it empirically from the database.

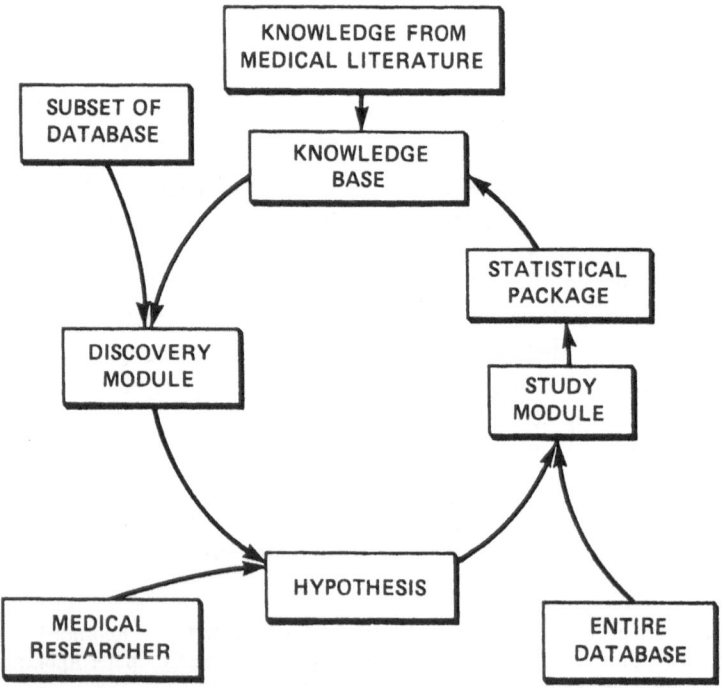

Figure 1. Discovery and confirmation in RX.

Following this concept, a prototype Discovery Module was added to the RX Project in 1980. The Discovery Module combs through a subset of the patient database to derive an ordered list of hypotheses for exploration. These hypotheses are studied by the Study Module as though they had been entered by a researcher. This cycle of discovery, confirmation, and incorporation is shown in Figure 1.

RX consists of five major parts: the Database, the Discovery Module, the Knowledge Base, the Study Module, and a statistical analysis package. A brief description of each follows.

The Time-Oriented Database

The database we use is the ARAMIS Database, the American Rheumatism Association Medical Information System, developed at Stanford University and implemented on TOD, a Time-Oriented Database system [5,6,7]. A recent review of clinical databases appears in [8]. Our research, so far, has been done entirely on a subset of the ARAMIS Database collected at the Stanford University Division of Immunology Clinics and containing the records of fifty patients with systemic lupus erythematosus.

Each patient's record consists of a matrix of values for a set of attributes that may be recorded each time the patient is seen in the clinic. Values for several hundred attributes can be recorded in ARAMIS. The attributes include signs, symptoms, lab tests, therapies, and indices of patient functional status. In general, the time intervals between clinic visits are not uniform, and patients are not on treatment protocols.

TOD is implemented in PL/1; ARAMIS is stored on an IBM 370/3033 computer at the Stanford University Center for Information Technology. On the other hand, the RX Project is implemented at two other computer facilities at Stanford University: SUMEX-AIM and SCORE. SUMEX-AIM features a DEC dual processor KI-10 running the TENEX operating system. SCORE has a DEC 20/60 running TOPS-20. Data transfer from ARAMIS is done by magnetic tape.

All RX computer programs are written in INTERLISP, a dialect of LISP, a language highly suited for knowledge manipulation. The RX source code with knowledge base comprises approximately 200 disk pages of 512 words of 36 bits each.

The Discovery Module

The Discovery Module produces hypotheses of the form "A causes B." The hypotheses denote that in a number of individual patient records "A precedes and is correlated with B." The current Discovery Module uses lagged nonparametric correlation across variables but within individual patient records. The p-values of the correlations across patients are then

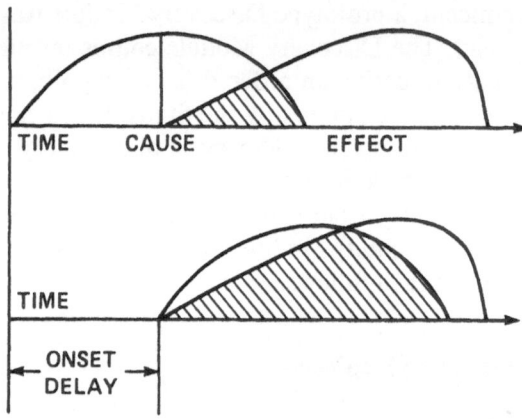

Figure 2. The principle underlying lagged correlation.

combined to yield a score that is used to order the list of hypotheses. Knowledge from the medical KB is used to determine the range of time lags examined. The principle underlying lagged correlation is shown in Figure 2.

The Knowledge Base

The leitmotif of the RX Project is that derivation of new knowledge from databases can best be performed by integrating existing knowledge of relevant parts of medicine and statistics into the medical information system. In the RX computer program the medical KB determines the operation of the Discovery Module, plays a pivotal role in the creation of subsequent studies in the Study Module, and finally serves as a repository for newly created knowledge. The medical KB grows by automatically incorporating new knowledge into itself. Hence, it is designed in such a way that relationships derived from the database are translated into the same machine-readable form as knowledge entered from the medical literature by clinicians.

The main data structure of RX's knowledge base (KB) is a tree representing a taxonomy of relevant aspects of medicine and statistics. Each object in the tree is represented as a schema containing an arbitrary number of property-value pairs. The RX KB contains approximately 250 schemata pertaining to medicine, 50 pertaining to statistics, and 50 system schemata. The medical knowledge in the RX KB covers only a small portion of what is known about systemic lupus erythematosus and limited areas of general internal medicine. The present KB is merely a test vehicle; its size is 50 disk pages or 120,000 bytes.

The most important class of properties in the schema corresponding to each medical object is that specifying the causal relationships of an object using an "effects" and an "affected-by" property list for each object. The resulting causal model is a directed cyclic graph; that is, the representation allows for the possibility that A causes B causes A with appropriate time lags.

Besides the simple fact that A may affect B, each causal relationship is represented by a set of features as below.

⟨INTENSITY, FREQUENCY, DIRECTION, SETTING
FUNCTIONAL FORM, VALIDITY, EVIDENCE⟩

Briefly, these take the following form when both the cause and effect are real valued.

- INTENSITY: the expected change in the effect given a change in the cause, expressed as an unstandardized regression coefficient,
- FREQUENCY: the distribution of the effect across patients, expressed as deciles of the expected effect given a "strong" change in the causal variable,
- DIRECTION: increase or decrease,
- SETTING: the clinical circumstances specifically included or excluded from the study, expressed as a Boolean with time-dependent predicates,
- FUNCTIONAL FORM: the complete statistical model used to study the relationship, expressed in machine-readable form,
- VALIDITY: a 1 to 10 scale distinguishing tentative associations from widely confirmed causal relationships,
- EVIDENCE: a summary of the study performed by the Study Module, including patient IDs, methods, and intermediate results.

The entire causal relationship is machine-readable. This enables it to be used automatically by the Study Module during subsequent studies. The causal relationships in the KB can also be interactively displayed in a variety of forms. All paths connecting two nodes may be displayed, or the details of a particular causal relationship: its mathematical form, the evidence supporting it, or its distribution across patients. In the example seen in Figure 3 the effects of prednisone have been displayed as translated from their internal representation.

PREDNISONE, at a level of 30 mg/day   {modal effects}

---

usually increases CHOLESTEROL by 50 to 130 mg/dl,
usually increases WEIGHT by 3 to 7 kg.
regularly attenuates NEPHROTIC SYNDROME by 1 to 2 g protein/24 hr.
regularly attenuates GLOMERULONEPHRITIS by 10 to 30% activity,
regularly decreases EOSINOPHILS by 2 to 3% of WBC,
commonly decreases ANTI-DNA by 50 to 90% activity,
occasionally increases GLUCOSE by 20 to 100 mg/dl.

Figure 3. Displaying the effects of Prednisone.

The Study Module

The Study Module is the core of the RX algorithm. It takes as input a causal hypothesis obtained either from the Discovery Module or interactively from a researcher. It then generates a medically and statistically plausible model of the hypothesis, which it analyzes on appropriate data from the database.

In creating a study design the Study Module follows accepted principles of epidemiological research. It determines study feasibility and study design: cross-sectional versus longitudinal. It uses the KB to determine the confounders of a given hypothesis, and it selects methods for controlling their influence: elimination of patient records, elimination of confounding time intervals, or statistical control. The Study Module then determines an appropriate statistical method using knowledge stored as production rules. Most studies have used a longitudinal design involving a multiple regression model applied to individual patient records. Results across patients are combined using weights based on the precision of the estimated regression coefficient for each patient. The steps in the Study Module appear below.

1. Parse the hypothesis and determine the classification of variables in it.
2. Determine the feasibility of the study on the database.
3. Select confounding variables and causal dominators using the KB.
4. Select methods for controlling the confounding variables.
5. Determine proxy variables.
6. Determine eligibility criteria.
7. Create a statistical model of the hypothesis using knowledge from the KB.
   a. Select an overall study design.
   b. Select statistical methods.
   c. Format the appropriate database access functions.
8. Run the study.
   a. Fetch the appropriate data from eligible patient records.
   b. Perform a statistical analysis of each patient's record.

IF   the number of patients affected by a variable is a small percentage of the number of patients in the study,

AND   the variable is present throughout those records,

THEN eliminate those records from the study.

Figure 4. A production rule used for determining the method of control.

Multiple regression

---

objectives: linear model

prerequisites:
  one dependent variable
  two or more independent variables
  measurement level of dependent variable = real valued
  measurement level of independent variable = real valued
  number of observations > 1 + number of independent variables

assumptions:
  independent and identically distributed errors
  normally distributed errors
  linear and additive effects

Figure 5. Frame corresponding to multiple regression.

    c. Combine the results to determine medical and statistical significance.
9. Incorporate the results into the knowledge base.

In this brief overview it is impossible to show these steps in detail. However, Figure 4 shows a typical production rule used in step 4.

Figure 5 shows some of the information in the schema for multiple regression that is used in step 7b.

Figure 6 shows a typical model that the Study Module assembles at step 7c.

The Statistical Package: IDL

Until July 1980, all statistical analyses were performed using SPSS as a subroutine. Currently all statistical analysis is done using IDL [9]. Writ-

---

$$\Delta \text{cholesterol} = \beta_0 + \beta_1 \, \Delta \text{albumin} + \beta_2 \, \Delta \log(\text{prednisone}),$$

where

$$\Delta \text{cholesterol} = \text{cholesterol}(t) - \text{cholesterol}(t_{\text{pchol}});$$
$$\Delta \text{albumin} = \text{albumin}(t - \tau_{\text{NS}}) - \text{albumin}(t_{\text{pchol}} - \tau_{\text{NS}});$$

and

$$\Delta \log(\text{prednisone}) = \log[\text{prednisone}(t - \tau_{\text{pred}})] - \log[\text{prednisone}(t_{\text{pchol}} - \tau_{\text{pred}})].$$

Figure 6. A regression model assembled by the study module.

ten in INTERLISP, IDL makes available fast numerical computation, matrix manipulation, and a variety of high-level primitives for statistical computation. To the basic IDL package we have added fifty disk pages of other statistical routines. The Study Module writes the study design to disk, then calls IDL. IDL reads the study design, executes it, writes the results to disk, then calls the Study Module.

The method of analysis we have used most often involved performing a separate multiple regression on each patient record, then combining results across patients. Our method of analysis accounts for autocorrelation and for differing quantities of data across patients.

Results, Availability, and Limitations

The current RX system was applied to a sample database containing the longitudinal records of 50 patients with systemic lupus erythematosus followed for an average of 50 clinic visits. Several well-known effects of the steroid drug prednisone were confirmed by the Study Module. The Study Module automatically incorporated these new links and details of the studies into the KB in the format previously discussed. Again, we hope that interested readers will examine the references [1] and [2] for details on methods and results.

The RX computer program is currently only a research prototype. It is not available outside our lab except for program development. Further development of the program is continuing in order to expand the repertoire of statistical models available and to generalize the syntax of causal relationships in the KB.

We must emphasize that any methodology that draws causal inferences based on nonrandomized data is subject to an important limitation: unknown covariates cannot be controlled. The strength of a particular knowledge base lies in its comprehensiveness, but even so, it cannot guarantee nonspuriousness. Only through repeated studies, particularly through experimental manipulation of the causal variable, can a given result become more definitive.

Acknowledgments. We are grateful to Guy Kraines, Kent Bailey, and Byron William Brown for their assistance with the statistical models, to Ronald Kaplan, and Beau Shiel for their assistance with IDL, and to James Fries, Dennis McShane, Alison Harlow, and James Standish for kindly providing access to the database.

Funding for this research was provided by the National Center for Health Services Research through grants HS-3650 and HS-4389, by the National Library of Medicine through grant LM-03370, and by the Pharmaceutical Manufacturers Association Foundation. Computation facilities were provided by SUMEX-AIM through NIH grant RR-00785 from the Biotechnology Resources Program. Clinical data were obtained from the American Rheumatism Association Medical Information System.

References

[1] Blum, Robert L.: Discovery and Representation of Causal Relationships from a Large Time-Oriented Clinical Database: The RX Project; Ph.D. Thesis, Stanford University, January, 1982, available as Stanford Computer Science Department Technical Report STAN-CS-82-900.

[2] Blum, Robert L.: Discovery, Confirmation, and Incorporation of Causal Relationships from a Large Time-Oriented Clinical Database: The RX Project, *Computers and Biomedical Research*, 15, 164–187, 1982.

[3] Blum, Robert L.: Displaying Clinical Data from a Time-Oriented Database; *Computers in Biology and Medicine*, 11:4, 1981.

[4] Blum, Robert L. and Wiederhold, G.: Inferring Knowledge from Clinical Data Banks Utilizing Techniques from Artificial Intelligence; Proc. 2nd Annual Symp. on Comp. Applic. in Med. Care, 303–307, IEEE, Washington, D.C., November 5–9, 1978.

[5] Weyl, S., Fries, J., Wederhold, G., Germano, F.: A Modular Self-Describing Clinical Databank System; *Computers and Biomedical Research*, 8:3, 279–293, June, 1975.

[6] Wiederhold, G. and Fries, J.F.: Structured Organization of Clinical Data Bases; AFIPS Conference Proceedings, 44: 479–485, 1975.

[7] Wiederhold, G.: Database Design. McGraw-Hill, 1977.

[8] Wiederhold, G.: Databases for Health Care; Springer-Verlag, 1981.

[9] Kaplan, R.M., et al: The Interactive Data-Analysis Language Reference Manual; Xerox Palo Alto Research Corp., 1978.

Explaining and Justifying Expert Consulting Programs

William R. Swartout

Traditional methods for explaining programs provide explanations by converting to English the code of the program or traces of the execution of that code. While such methods can provide adequate explanations of what the program does or did, they typically cannot provide justifications of the code without resorting to canned-text explanations. That is, such systems cannot tell why what the system is doing is a reasonable thing to be doing. The problem is that the knowledge required to provide these justifications is needed only when the program is being written and does not appear in the code itself.

The XPLAIN system uses an automatic programmer to generate the consulting program by refinement from abstract goals. The automatic programmer uses a domain model, consisting of facts about the application domain, and a set of domain principles which drive the refinement process forward. By examining the refinement structure created by the automatic programmer it is possible to provide justifications of the code. This paper discusses the system described above and outlines additional advantages this approach has for explanation.

1. Introduction

To be acceptable, expert programs must be able to explain what they do and justify their actions in terms understandable to the user. Expert programs usually have some heuristic basis. While these heuristics may provide good performance for most cases, there may be unusual cases where they produce erroneous results, or where the rationale for using them is faulty. If a user is suspicious of the advice he receives, he should be able

Reprinted from the *Proceedings of the Seventh International Joint Conference on Artificial Intelligence,* pp. 815–822, 1981. Used by permission of the International Joint Conferences on Artificial Intelligence, Inc.; copies of the Proceedings are available from William Kaufmann, Inc., 95 First Street, Los Altos, CA 94022, U.S.A.

to ask for a decision of the methods employed and the reasons for employing them. In addition, the scope of expert systems, like that of human experts, is often quite narrow. An expansion facility can help a user discover when a system is being pushed beyond the bounds of its expertise.

In the area of medical consultant programs[1] the need for explanation is particularly acute. In designing a consultant program, we must consider what sorts of capabilities we are trying to provide for the physician-user. If we consider the interaction between a physician and a human consultant, we realize that it is not just a simple one-way exchange where the physician provides data and the consultant provides an answer in the form of a prescription or diagnosis. Rather, there is typically a lively dialog between the two. The physician may question whether some factor was considered or what effect a particular finding had on the final outcome. Viewed in this light, we realize that a computer program which only collects data and provides a final answer will not be found acceptable by most physicians. In addition to providing diagnoses or prescriptions, a consultant program must be able to explain what it's doing and justify why it's doing it.

Researchers have recognized this, and many proposals for new expert systems have at least mentioned the need for explanation. Some systems have actually provided an explanatory facility. Yet existing approaches to explanation fail in some important ways. This paper will document these failings and describe an approach toward their solution.

While we have concentrated on the problem of providing explanations to medical personnel, we do not feel that the need for explanation is limited to medicine nor that the techniques we have developed for explanation and justification are limited to medical applications. Medical programs provide a good testbed for the general problem of explaining a consulting program to the audience it is intended to serve.

The next section will describe the Digitalis Therapy Advisor, the program we have chosen as a testbed for our ideas about explanation, and some of the medical aspects of digitalis therapy. After that, we will describe some of the problems with previous explanation systems and the approach we have taken to overcome those problems.

2. Digitalis Therapy and the Digitalis Advisor

The digitalis glycosides are a group of drugs that were originally derived from the foxglove, a common flowering plant. Their principal effect is to

---

[1]Some medical consultant programs include: MYCIN—a program that aids physicians with antimicrobial therapy [Shortliffe76]. INTERNIST—a program that makes diagnoses in internal medicine [Pople77] and PIP—a program that makes diagnoses primarily in the area of renal disease [Pauker76].

strengthen and stabilize the heartbeat. In current practice, digitalis is prescribed chiefly to patients who show signs of congestive heart failure
(CHF) and/or conduction disturbances of the heart. Congestive heart failure refers to the inability of the heart to provide the body with an adequate blood flow. This condition causes fluid to accumulate in the lungs
and outer extremities and it is this aspect that gives rise to the term
"congestive". Digitalis is useful in treating this condition, because it
increases the contractility of the heart, making it a more effective pump.
A conduction disturbance appears as an arrhythmia, which is an unsteady
or abnormally paced heartbeat. Digitalis tends to slow the conduction of
electrical impulses through the conduction system of the heart, and thus
steady certain types of arrhythmias. Due to the positive effect that digitalis has on the heart, it is one of the most commonly used drugs in the
United States.

Like many other drugs, digitalis can also be a poison if too much is
administered. For a variety of reasons, including a small therapeutic win-
dow, subtle signs of toxicity and high interpatient variability, digitalis is
difficult to administer. One complication the physician must deal with is
the possibility that his patient may be more sensitive to the drug (for
whatever reason) than the average patient. If a physician knows those
factors that make a patient more sensitive he can reduce the likelihood
of overdosing (or underdosing) the patient by adjusting the dose depending on whether he observes the sensitizing factors or not.

Over the years, a number of factors have been identified that increase
the automaticity of the heart.[2] These include: a low level of serum potassium (hypokalemia), a high level of serum calcium (hypercalcemia), damage to the heart muscle (cardiomyopathy), and a recent myocardial
infarction (among others). When these exist in conjunction with digitalis
administration, the automaticity can be increased substantially. We will
concentrate on just the first three in this paper.[3]

2.1 The Digitalis Therapy Advisor Testbed

A few years ago, a Digitalis Therapy Advisor was developed at MIT by
Pauker, Silverman, and Gorry [Silverman75, Gorry78]. This program

---

[2]In the normal heart, there is a place in the left atrium called the sino-atrial (SA) node, which
sets the pace for the heart. Under the right circumstances, other parts of the heart can take
over the pace-setting function. Sometimes this can be life-saving if, for example, the SA
node is damaged. But at other times it can be life-threatening, since several pace-makers
operating simultaneously tend to increase the likelihood of setting up a dangerous arrhythmia. When we say that digitalis increases the automaticity of the heart, we mean that digitalis increases the tendency of other parts of the heart to take over the pace-setting function
from the SA node.
[3]The XPLAIN system currently only knows about the first three factors, although it would
not be particularly difficult to expand it to cover the others.

was later revised and given a preliminary explanatory capability [Swartout77]. The limitations of these explanations (and of those produced by similar techniques) will be discussed below. This program differed from earlier digitalis advisors [Peck73, Jelliffe70, Jelliffe72, Sheiner72] in two important respects. First, when formulating dosage schedules, it anticipated possible toxicity by taking into account the factors that increased digitalis sensitivity and reduced the dose when those factors were present. Second, the program made assessments of the toxic and therapeutic effects which actually occurred in the patient after receiving digitalis to formulate subsequent dosage recommendations. This program worked in an interactive fashion. The program would ask the physician for data about the patient and produce recommendations after that data was entered. When the dose of digitalis was being adjusted, the physician was asked to consult with the program again to assess the patient's response. This is the program we used as a testbed for our work in explanation and justification. In the remainder of the paper, we will refer to this program as the "old Digitalis Advisor".

3. Kinds of Questions

In the spring of 1979, we conducted a series of informal trials in an attempt to discover what kinds of questions occurred to medical personnel as they ran the old Digitalis Advisor. In their trial, medical students and fellows were asked to run the program and ask questions (verbally) as they occurred to them. The author attempted to answer these questions. The interactions were tape recorded and later transcribed.

No formal analysis of this data was attempted, but examination of the transcripts did provide an indication of types of questions that might arise while running a consulting program.

These included:

1. Questions about the methods the program employed:

User: "How do you calculate your body store goal? That's a little lower than I anticipated."

This sort of question could be answered by the explanation routines of the old Digitalis Advisor. It can also be answered by the system presented in this paper.

2. Justifications of the program's actions:

User: (peruses recommendations) "Why do we want to make a temporary reduction?"

Author: "We're anticipating surgery coming up and surgery, even non-cardiac surgery, can cause increased sensitivity to digitalis, so it wants to temporarily reduce the level of digitalis."

This is exactly the sort of question we are concentrating on in this paper.

3. Questions involving confusion about the meaning of terms:

IS THE RENAL FUNCTION STABLE?
THE POSSIBILITIES ARE:
   1. STABLE
   2. UNSTABLE
ENTER SINGLE VALUE = = = = >

User: "Now this question . . . I'm not really sure . . . 'renal function stable'
does it mean stable abnormally or . . . because I mean, the patient's renal
function is not normal but it's stable at the present time."

Author: "That's what it means."

This paper will not address this last type of question.

4. Previous Approaches to Explanation

A number of different approaches have been taken to attempt to provide
programs with an explanatory capability. The major approaches include
using (1) previously prepared text to provide explanations and (2) producing explanations directly from the computer code and traces of its
execution.

The simplest way to get a computer to answer questions about what it
is doing is to anticipate the questions and store the answers as English
text. Only the text that has been stored can be displayed. This is called
canned text, and explanations produced by displaying canned text are
called canned explanations. The simplest sorts of canned explanations
are error messages. For example, a medical program designed to treat
adults might print the following message if someone tried to use it to treat
an infant:

THE PATIENT IS TOO YOUNG TO BE TREATED BY THIS
PROGRAM.

It is relatively easy to get a small program to provide English explanations
of its activity using this canned text approach. After the program is written, canned text is associated with each part of the program explaining
what that part of the program is doing. When the user wants to know
what's going on, the computer merely displays the text associated with
what it's doing at the moment.

There are several problems with the canned text approach to explanation. The fact that the program code and the text strings that explain
that code can be changed independently makes it difficult to guarantee
consistency between what the program does and what it claims to do.
Another problem with the canned text approach is that all questions and

answers must be anticipated in advance and the programmer must provide answers for all the questions that the user might ask. For large systems, that is a nearly impossible task. Finally, the system has no conceptual model of what it is saying. That is, to the computer, one text string looks much like any other, regardless of the content of that string. Thus, it is difficult to use this approach if we want our system to provide more advanced sorts of explanations such as suggesting analogies or giving explanations at different levels of abstraction.

Another approach to explanation is to produce explanations directly from the program [Davis76, Shortliffe76, Swartout77, Winograd71]. That is, the explanation routines examine the program that is executed. Then by performing relatively simple transformations on the code these explanation routines can produce explanations of how the system does things. For example, the old Digitalis Advisor could examine the code it used to check for increased digitals sensitivity caused by increased serum calcium and produce an explanation of what that code did (as shown in Figure 1).

The old Digitalis Advisor, like most similar systems, also maintained an execution trace. The trace can be examined by the explanation routines to tell what the system did for a particular patient. Figure 2 describes how the system checked for myxedema. The system also had a limited ability to explain why it asked the user a question. Figure 3 shows the system's response when the user wanted to know why he was being asked about serum calcium.

Since the explanation routines only perform simple transformations on the program code, the quality of the explanations produced in this manner depends to a great degree on how the system code is written. In particular, the basic structure of the program is not altered significantly, and the names of variables in the explanation are basically the same as

---

TO CHECK SENSITIVITY DUE TO CALCIUM I DO THE FOLLOWING STEPS:

1. I DO ONE OF THE FOLLOWING:

1.1 IF EITHER THE LEVEL OF SERUM CALCIUM IS GREATER THAN 10 OR INTRAVENOUS CALCIUM IS GIVEN THEN I DO THE FOLLOWING SUBSTEPS:

1.1.1 I SET THE FACTOR OF REDUCTION DUE TO HYPERCALCEMIA TO 0.75.

1.1.2 I ADD HYPERCALCEMIA TO THE REASONS OF REDUCTION.

1.2 OTHERWISE, I REMOVE HYPERCALCEMIA FROM THE REASONS OF REDUCTION AND SET THE FACTOR OF REDUCTION DUE TO HYPERCALCEMIA TO 1.00.

Figure 1. Explanation of how the old digitalis advisor checked hypercalcemia.

I CHECKED SENSITIVITY DUE TO THYROID-FUNCTION BY EXECUTING THE FOLLOW-
ING STEPS:

1. I ASKED THE USER THE STATUS OF MYXEDEMA. THE USER RESPONDED THAT
THE STATUS OF MYXEDEMA WAS PRESENT.

2. SINCE THE STATUS OF MYXEDEMA WAS PRESENT I DID THE FOLLOWING:

2.1 I ADDED MYXEDEMA TO THE PRESENT AND CORRECTABLE CONDITIONS. THE
PRESENT AND CORRECTABLE CONDITIONS THEN BECAME MYXEDEMA.

2.2 I REMOVED MYXEDEMA FROM THE DEGRADABLE CONDITIONS. THE DEGRAD-
ABLE CONDITIONS THEN BECAME HYPOKALEMIA, HYPOXEMIA, CARDIOMYOPA-
THIES-MI, AND POTENTIAL POTASSIUM LOSS DUE TO DIURETICS.

2.3 I SET THE FACTOR OF REDUCTION DUE TO MYXEDEMA TO 0.67. THE FACTOR
OF REDUCTION DUE TO MYXEDEMA WAS PREVIOUSLY UNDETERMINED.

2.4 I ADDED MYXEDEMA TO THE REASONS OF REDUCTION. THE REASONS OF
REDUCTION THEN BECAME MYXEDEMA.

Figure 2. Explaining how thyroid function was checked.

those in the program. If the explanations are to be understandable, the
expert system must be written so that its structure is easily understood
by anyone familiar with its domain of expertise, and the variable and
procedure names used in the program must represent concepts which are
meaningful to the user.

This method of producing explanations has some advantages. It is relatively simple. If the right way of structuring the problem can be found,
it does not impose too great a burden on the programmer; since the explanations reflect the code directly, consistency between explanation and
code is assured.

Despite these advantages, there are some serious problems with this
technique. It may be difficult or impossible to structure the program so

WHAT IS THE LEVEL OF SERUM CALCIUM?

ENTER NUMERICAL VALUE = = = = > why? *(The user types "why?" indicating that
he questions why he is being asked to supply the value of serum calcium.)*

MY TOP GOAL IS TO BEGIN THERAPY. ONE STEP IN DOING THAT IS TO CHECK SEN-
SITIVITIES. I AM NOW TRYING TO CHECK SENSITIVITY DUE TO CALCIUM.

Figure 3. A limited explanation telling why a question is asked.

that the user can easily understand it. The fact that every operation performed by the computer must be explicitly spelled out sometimes forces the programmer to program operations which a physician would perform without thinking about them. That problem is illustrated in Figure 2. Steps 2.1, 2.2, and 2.4 are somewhat mystifying. In fact, these steps are needed by the system so that it can record what sensitivities the patient had that made him more likely to develop digitalis toxicity. These steps are involved more with record keeping than with medical reasoning, but they must appear in the code so that the computer will remember why it made a reduction. Since they appear in the code, they are described by the explanation routines, although they are more likely to confuse a physician-user than enlighten him. An additional problem is that it is difficult to get an overview of what is really going on here. While the system is explicit about record keeping, it is not very explicit about the fact that it is going to reduce the dose, though it hints at a reduction by saying that the "factor of reduction" is being set to 0.67.

An additional problem, and the primary one we will address in this paper, is that while this way of giving explanations can state *what* the system does or did, it has only a limited ability to state why the system did what it did (see Figure 3). That is, the system can't give adequate justifications for its actions. In the explanations given above, the system can't state that it reduces the dose because increased calcium causes increased automaticity. The information needed to justify the program is the information that was used by the programmer to write the program, but it does not have to be incorporated into the program for the program to perform successfully—just as one can successfully bake a cake without knowing why baking powder appears in the recipe. Since it is desirable for expert programs to be able to justify what they do as well as do it successfully, we need to find a way of capturing the knowledge and decisions that went into writing the program in the first place. The remainder of this report will describe recent efforts we have made toward achieving that goal in the context of the Digitalis Therapy Advisor.[4]

5. Providing Justifications

We need a way of capturing the knowledge and decisions that went into writing the program. One way to do this is to give the computer enough knowledge so that is can write the program itself and remember what it did. This notion of automatic programming has been researched considerably [Balzer77, Barstow77, Green79, Long77, Manna77] and is not a new idea. Using an automatic programmer to help in producing expla-

---

[4][Clancey79] notes that even in rule-based systems, knowledge is often too "compiled" resulting in explanation problems very similar to the ones described here.

nations is a new idea. Since we are primarily interested in explanation, we have chosen not to deal with a number of problems that arise in automatic programming, including: choosing between different implementations, backup and recovery from dead-end refinements, and optimization.

5.1 System Overview

An overview of the XPLAIN system is given in Figure 4. The system has five parts: a Writer, a Domain Model, a set of Domain Principles, an English Generator, and a generated Refinement Structure. The Writer is an automatic programmer. It wrote new code which captured the functionality of major portions of the old Digitalis Advisor.[5] The Domain Model and the Domain Principles contain knowledge about the domain of expertise. Thus, in this case, they contain information about digitalis and digitalis therapy. They provide the Writer with the knowledge it needs to write the code for the Digitalis Advisor. The Refinement Structure can be thought of as a trace left behind by the Writer. It shows how the Writer develops the Digitalis Advisor. When a physician-user runs the Digitalis Therapy advisor, he can ask the system to justify why the program is doing what it is doing. The Generator gives him an answer by examining the Refinement Structure and the step of the Advisor currently being executed. If we wanted to write a new program covering a new medical domain, we would have to change the Domain Model and the Domain Principles, but we should not have to change the Writer or the English Generator.[6]

The Refinement Structure is created by the Writer from the top level goal (in this case Administer Digitalis) as it writes the Digitalis Advisor. The Refinement Structure is a tree of goals, each being a refinement of the one above it in the tree (see Figure 5). By "refining a goal" we mean taking a goal and turning it into more specific subgoals. Looking at Figure 5, we see that the top of the tree is a very abstract goal, in this case, Administer Digitalis. This goal is refined into less abstract steps by the Writer. These more specific steps are steps the system executes to administer digitalis. For example, one such step is to Anticipate Toxicity, that is, to anticipate whether the patient may become toxic due to increased digitalis sensitivity. The Writer then refines this more specific goal to a still more specific goal. Eventually, the level of system primitives is reached. System primitives are operations which are built-in. Normally

---

[5]The code that has been written includes code to anticipate toxicities and to check for and assess various types of toxicities that may occur. As is discussed in [Swartout81], it should not be too difficult to complete the remainder of the implementation so that the functionality of the old Digitalis Therapy Advisor is completely captured.
[6]Note that the Writer writes the program once, and once written, the program is static. It is not written "on the fly" during interaction with the user/physician.

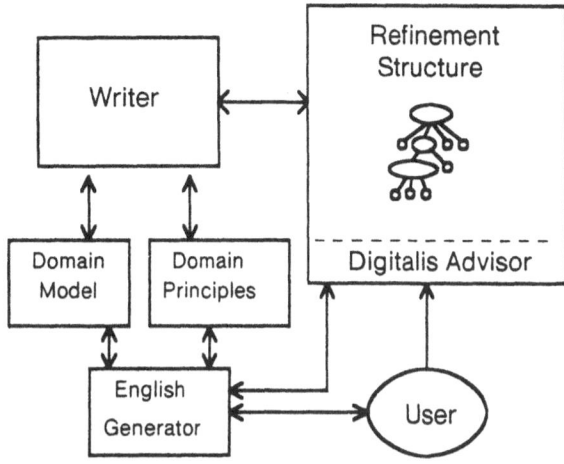

Figure 4. System overview.

they are very basic, simple operations, so the fact that they cannot be explained is usually not a problem. Typical primitives include those that perform arithmetic operations like PLUS and TIMES and those that set variables to a particular value. The leaves of the refinement structure constitute the basic operations performed by the Digitalis Advisor, the program that we wanted the automatic programmer to produce.

The Domain Model is a model of the facts of the domain. In this case,

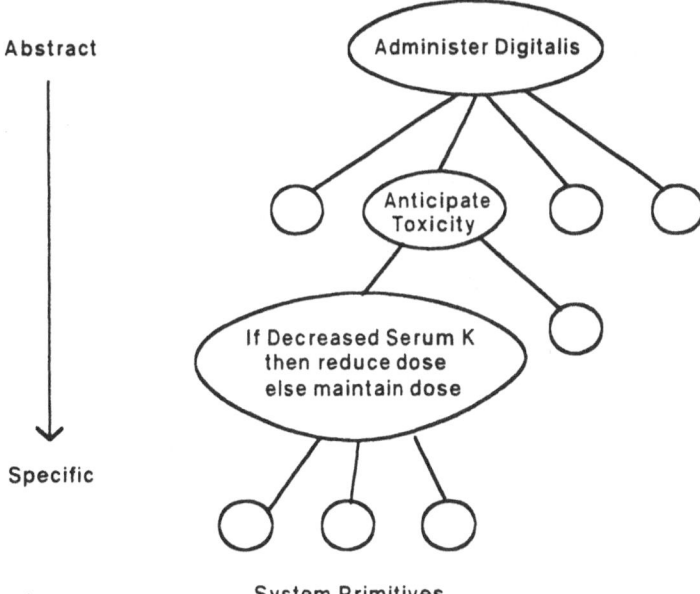

Figure 5. A sample Refinement Structure.

it is a model of the causal relationships in digitalis therapy. A simplified portion of the model is shown in Figure 6. In this model, the boxes are states, and the arrows represent causality. This model shows some of the effects of increased digitalis. It also shows that increased serum Ca and decreased serum K can each cause increased automaticity. These facts correspond to the sorts of facts that a medical student learns in class during the first two years of medical school. These facts are static. That is, they have no notion of process. The model says that increased digitalis can cause a change to ventricular fibrillation but it doesn't say what to do about it. Medical students go to medical school for an additional two years and acquire these procedures by observing more experienced personnel as they practice medicine on the wards. The set of Domain Principles provides the Writer with this sort of problem-solving knowledge.

Domain Principles tell the Writer how something (such as prescribing a drug or analyzing symptoms) should be done. They guide it as it refines abstract goals to more specific ones. A (somewhat simplified) Domain Principle appears in Figure 7.[7] This particular Principle helps the writer in anticipating digitalis toxicity. It represents the common sense notion that if one is considering administering a drug, and there is some factor that enhances the deleterious effects of that drug, then if that factor is present in the patient, less drug should be given. This Principle has three parts: a Goal, a Domain Rationale, and a Prototype Method.

The goal tells the Writer what it is that the Principle can help it do. In this case, the Principle can help the Writer in anticipating toxicity. The Domain Rationale is a pattern which is matched against the Domain Model to see where it is appropriate to achieve the goal. In the example, the system will look in the Domain Model to match a **finding** (e.g. increased Ca) which causes some sort of a **dangerous deviation** (e.g. change to ventricular fibrillation) which is also caused by an increased level of the drug. By looking at the Domain Model, we can see both increased Ca and decreased K will match as findings, since both can cause a change to ventricular fibrillation.

The Prototype Method is an abstract method that tells the system how to accomplish the goal. The steps of the Prototype Method are annotated to distinguish implementation details (such as record-keeping) from steps which are significant in medical problem-solving. These annotations are used by the explanation routines to filter out implementation details when presenting explanations to medical personnel.

After the domain rationale has been matched against the domain model, the Prototype Method is instantiated for each match of the

---

[7]Domain Principles are composed of variables and constants. Variables appear in boldface in Figure 7. When the Writer is matching, a variable in a pattern will match anything which is of the same kind as itself. Thus, the variable finding would match increased serum-Ca or decreased K, since increased serum-Ca and decreased K are both kinds of findings.

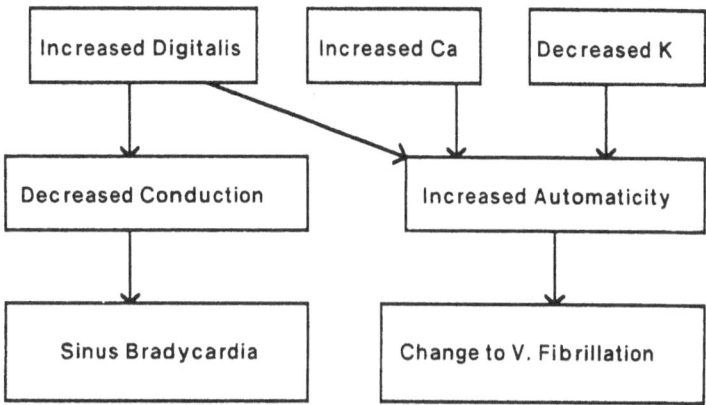

Figure 6. A simplified portion of the Domain Model.

Domain Rationale. When we say that we instantiate the Prototype Method, that means that we create a new structure where the variables in the Prototype Method have been replaced by the things they matched. In this case, two structures would be created. In the first, **finding** would be replaced by increased serum Ca and **drug** would be replaced by digitalis. In the second, **finding** would be replaced by decreased serum K and **drug** would again be replaced by digitalis. Note that now, with these new structures, we have changed the single abstract problem of how to anticipate toxicity into several more specific ones, such as how to determine whether decreased serum K exists, how to reduce the dose, and how to maintain it.

After instantiation, the more specific goals of the Prototype Method

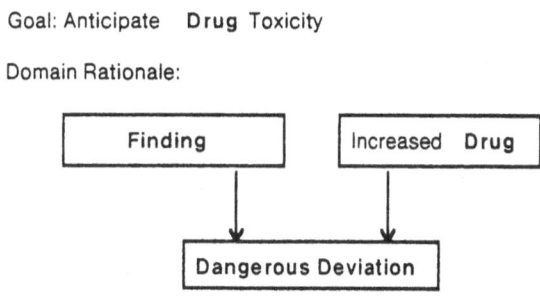

Figure 7. An example of a Domain Principle.

are placed in the Refinement Structure as sons of the goal being resolved. If we look at Figure 5, we can see that the instantiated Prototype Method that checks for decreased serum K has been placed below the Anticipate Toxicity goal. Once they have been placed in the Refinement Structure, the newly instantiated goals become goals for the Writer to resolve. For example, after the Writer applied this Domain Principle, it would have to find ways of determining whether increased calcium existed in the patient, whether decreased potassium existed, and ways of reducing and maintaining the dose. The system continues in this fashion, refining goals at the bottom of the structure and growing the tree down and down until eventually the level of system primitives is reached.

The system must also take into account interactions between the actions it takes. For example, the individual instantiations above say that if increased serum calcium exists the dose should be reduced, and if decreased serum potassium exists the dose should be reduced, but they don't give any indication of what should happen if *both* increased calcium *and* decreased serum potassium co-occur. Exactly what should happen depends on the characteristics of the domain. It could be that the occurance of either sensitivity "covers" for the other, so that only one reduction should be made and the predicate of the IF should be made into a disjunction. Or (as is actually the case), it could be that when multiple sensitivities appear, multiple reductions should be made. The way

---

Please enter the value of serum-k: why?

The system is anticipating digitalis toxicity. Decreased serum-k causes increased automaticity, which may cause a change to ventricular fibrillation. Increased digitalis also causes increased automaticity. Thus, if the system observes decreased serum-k, it reduces the dose of digitalis due to decreased serum-k.

Please enter the value of serum-k: 3.7

Please enter the value of serum-ca: why?

(The system produces a shortened explanation, reflecting the fact that it has already explained several of the causal relationships in the previous explanation. Also, since the system remembers that it has already told the user about serum-K, it suggests the analogy between the two here.)

The system is anticipating digitalis toxicity. Increased serum-ca also causes increased automaticity. Thus, (as with decreased serum-k) if the system observes increased serum-ca, it reduces the dose of digitalis due to increased serum-ca.

Please enter the value of serum-ca: 9

Figure 8. A sample interaction providing justifications.

to resolve that is to serialize these two program fragments, connecting the outputs of the first to the inputs of the second. The automatic programmer handles this situation by setting it up as something to be refined. The Domain Principle used in the refinement of this problem may further constrain the way in which other goals may be refined. The details of this operation will not be presented here. The interested reader should see [Swartout81].

Once the refinement process is complete, we have a working expert system. A sample interaction with the system is given in Figure 8. The first sentence of the explanation was produced by stating the higher goal (that is, anticipate toxicity). Next, the explanation routines located the Domain Principle which caused the step in question to appear in the program. The Domain Rationale associated with that Principle was then converted to English (with pattern variables replaced by the facts in the Domain Model they matched). That step produced the next two sentences of the explanation. The last sentence is just the instantiated version of the Prototype Method of the Domain Principle. These explanations should be compared with those presented in Figure 3 to appreciate the improvement that is possible with this approach. (The generation routines are described in detail in [Swartout81].)

5.2 Explanations of Domain Principles

In the original version of the Digitalis Advisor, when we wanted to give a more abstract view of what was going on, we just described a higher level procedure [Swartout77]. In this regard, we were following the principles of structured programming. While this approach often gave reasonable explanations, there were times when it was considerably less than illuminating. The general method for anticipating digitalis toxicity was called "Check Sensitivities" in the old version of the Digitalis Advisor. An explanation of it appears in Figure 9. While this explanation does tell the user what sensitivities are being checked,[8] it does not say what will be done if sensitivities are discovered nor does it say why the system considers these particular factors to be sensitivities. Finally, it is much too redundant and verbose. The first objection can be dealt with by removing the calls to lower procedures and substituting the code of those procedures in-line. This results in the somewhat improved explanation produced by XPLAIN when it is asked to describe the method for anticipating digitalis toxicity (see Figure 10). However, while this explanation

---

[8]The reader may notice that there were more sensitivities checked in the original version of the program than in the current version. We now feel that some of these, such as thyroid function and advanced age, should not be treated as sensitivities per se because they tend to have an effect on reducing renal function and hence slowing excretion, rather than on increasing sensitivity to digitalis. The other sensitivities would be easy to add by including the appropriate causal links in the domain model.

```
(describe-method [(check sensitivities)])
```

TO CHECK SENSITIVITIES I DO THE FOLLOWING STEPS:
1. I CHECK SENSITIVITY DUE TO CALCIUM.
2. I CHECK SENSITIVITY DUE TO POTASSIUM.
3. I CHECK SENSITIVITY DUE TO CARDIOMYOPATHY-MI.
4. I CHECK SENSITIVITY DUE TO HYPOXEMIA.
5. I CHECK SENSITIVITY DUE TO THYROID-FUNCTION.
6. I CHECK SENSITIVITY DUE TO ADVANCED AGE.
7. I COMPUTE THE FACTOR OF ALTERATION.

Figure 9. An explanation from the old digitalis therapy advisor.

shows what the system does, it doesn't say why things like increased calcium, cardiomyopathy and decreased potassium are sensitivities, and if anything, it is even more verbose than the original explanation.

The reason we can't get the sorts of explanations we want by producing explanations directly from the code is that much of the sort of reasoning we want to explain has been "compiled out." Thus, we are forced into explaining at a level that is either too abstract or too specific. The intermediate reasoning which we would like to explain was done by a human programmer in the case of the old Digitalis Advisor. However, because this performance program was produced by an automatic programmer, that reasoning is available in the Domain Principle. For example, if we were to use the English generator to explain the Domain Principle that produced the code for anticipating digitalis toxicity rather than the code itself we would get the explanation that appears in Figure 11. Thus, the use of an automatic programmer not only allows us to justify the performance program, but it also allows us to give better descriptions of methods by making available intermediate levels of abstraction which were not previously available.

```
(describe-method [((ANTICIPATE*O(TOXICITY*F DIGITALIS))*| 1)])
```

To anticipate digitalis toxicity:
1. If the system determines that cardiomyopathy exists, it reduces the dose of digitalis due to cardiomyopathy.
2. If the system determines that decreased serum-k exists, it reduces the dose of digitalis due to decreased serum-k.
3. If the system determines that increased serum-ca exists, it reduces the dose of digitalis due to increased serum-ca.

Figure 10. An explanation from the code for anticipating toxicity.

(describe-proto-method [(anticipate*o(toxicity*f digitalis))])

The system considers those cases where a finding causes a dangerous deviation and increased digitalis also causes the dangerous deviation. If the system determines that the finding exists, it reduces the dose of digitalis due to the finding.

The findings considered are increased calcium, decreased postassium and cardiomyopathy.

Figure 11. Explanation of a domain principle.

6. Is Automatic Programming too Hard?

One possible objection to the whole approach to explanation advocated in this paper is that it is just too hard to get an automatic programmer to write the performance program. When I first began this research, I thought that was the case. The original plan for producing better explanations was to create structures detailing the development of the performance program, but these structures would be created by hand rather than automatically. It was feared that automatic programming was just too hard. However, as the research progressed, it became clear that if we had sufficiently powerful representations available so that it could be said that explanations were being produced from an understanding of the program, then actually writing the program automatically wouldn't be all that much more difficult. I suspect this is true in general. It seems that the primary difficulty in both explanation and automatic programming is a knowledge representation problem, and that the kinds of knowledge to be represented in both cases are similar so that a solution to one case makes the other much easier. However, it must be pointed out that the field of automatic programming is still an active research area and a number of difficult problems remain to be solved in addition to the knowledge representation problem so this conjecture still awaits a final resolution.

7. A Summary of Major Points

First, we have argued that to be acceptable, consultant programs must be able to explain what they do and why. Second, we have described the various ways that traditional approaches fail to provide adequate explanations and justifications. Major failings include: (1) the inability of such approaches to justify what the system is doing because the knowledge required to produce justifications is not represented within the system, and (2) a lack of distinction between steps required just to get the computer-based implementation to work, and those that are motivated by the application domain. Third, we have outlined an approach which captures the knowledge necessary to improve explanations. This involves using an automatic programmer to generate the performance program. As the program is generated, a refinement structure is created which gives the explanation routines access to decisions made during the creation of the program. The improvement in explanatory capabilities that is achieved is due more to the availability of this refinement structure than to the use of more sophisticated English generation functions, since the explanation routines used in this paper do not differ greatly from those used in the old Digitalis Advisor.

Acknowledgments. This research was supported (in part) by the National Institutes of Health Grant No. 1 PO1 LM 03374-01 from the National Library of Medicine. The author wishes to express his thanks to Peter Szolovits for his insightful comments and suggestions during the course of this research.

References

[Barstow77] Barstow, D., "A Knowledge-Based System for Automatic Program Construction," Proceedings of the Fifth International Conference on Artificial Intelligence, 1977

[Balzer77] Balzer, R., Goldman, N., Wile, D., "Informality in Program Specifications," Proceedings of the Fifth International Conference on Artificial Intelligence, 1977

[Clancey79] Clancey, W.J., "Transfer of Rule-based Expertise Through a Tutorial Dialogue," Stanford University, Department of Computer Science, STAN-CS-79-769, 1979

[Davis76] Davis, R., "Applications of Meta Level Knowledge to the Construction, Maintenance and Use of Large Knowledge Bases," PhD thesis, Stanford Artificial Intelligence Laboratory Memo 283(1976).

[Gorry78] Gorry, G.A., Silverman, H., and Pauker, S.G., Capturing Clinical Expertise: A Computer Program that Considers Clinical Responses to Digitalis, *American Journal of Medicine* **64**:452–460, (March 1978).

[Green79] Green, C.C., Gabriel, R.P., Kant, E., Kedzierski, B.I., McCune, B.P., Phillips, J.V., Tappel, S.T., Westfold, S.J., "Results in Knowledge Based Program Synthesis," Proceedings of the Sixth International Joint Conference on Artificial Intelligence, 1979

[Jelliffe70] Jelliffe R.W., Buell J., Kalaba R. et al, "A Computer Program for Digitalis Dosage Regimens," *Math. Biosci.* 9:179–193, 1970

[Jelliffe72] Jelliffe R.W., Buell J, Kalaba R, "Reduction of digitalis toxicity by computer-assisted glycoside dosage regimens," *Ann. Intern. Med.* 77:891–906, 1972

[Long77] Long, W.J., "A Program Writer," MIT Laboratory for Computer Science, TR-187, 1977

[Manna77] Manna, Z., Waldinger, R., "The Automatic Synthesis of Systems of Recursive Programs," Proceedings of the Fifth International Conference on Artificial Intelligence, 1977

[Peck73] Peck C.C., Sheiner L.B. et al: "Computer-assisted Digoxin Therapy," *New England Journal of Medicine* 289:441–446, 1973.

[Pauker76] Pauker, S.G., Gorry, G.A., Kassirer, J.P., and Schwartz. W.B., "Toward the Simulation of Clinical Cognition: Taking a Present Illness by Computer," *The American Journal of Medicine* 60:981–995 (June 1976).

[Pople77] Pople, H.E., Jr., "The Formation of Composite Hypotheses in Diagnostic Problem Solving: an Exercise in Synthetic Reasoning," Proceedings of the Fifth International Joint Conference on Artificial Intelligence (1977).

[Sheiner72] Sheiner L.B., Rosenberg B., Melmon K., "Modelling of Individual Pharmacokinetics for Computer-aided Drug Dosage," *Computers and Biomedical Research* 5:441–459, 1972

[Shortliffe76] Shortliffe, E.H., *Computer Based Medical Consultations: MYCIN,* Elsevier North Holland Inc. (1976)

[Silverman75] Silverman, H., "A Digitalis Therapy Advisor," MIT Project MAC TR-143, 1975

[Swartout77] Swartout, W.R., "A Digitalis Therapy Advisor with Explanations," *Proceedings of the Fifth International Joint Conference on Artificial Intelligence,* August 1977

[Swartout81] Swartout, W.R., "Producing Explanations and Justifications of Expert Consulting Systems," MIT Laboratory for Computer Science Technical Report (in press)

[Winograd71] Winograd, T., "A Computer Program for Understanding Natural Language," MIT Artificial Intelligence Laboratory TR-17, 1971

Causal Understanding of Patient Illness in Medical Diagnosis

Ramesh S. Patil, Peter Szolovits, and William B. Schwartz

First generation *AI in Medicine* programs have clearly demonstrated the usefulness of AI techniques. However, it has also been recognized that the use of notions such as causal relationships, temporal patterns, and aggregate disease categories in these programs has been too weak. From our study of clinician's behavior we realized that a diagnostic or therapeutic program must consider a case at various levels of detail to integrate overall understanding with detailed knowledge. To explore these issues, we have undertaken a study of the problem of providing expert consultation for electrolyte and acid-base disturbances. We have partly completed an implementation of ABEL, the diagnostic component of the overall effort. In this paper we concentrate on ABEL's mechanism for describing a patient. Called the *patient-specific model,* this description includes data about the patient as well as the program's hypothetical interpretations of these data in a multi-level causal network. The lowest level of this description consists of pathophysiological knowledge about the patient, which is successively aggregated into higher level concepts and relations, gradually shifting the content from pathophysiological to syndromic knowledge. The aggregate level of this description summarizes the patient data providing a global perspective for efficient exploration of the diagnostic possibilities. The pathophysiological level description provides the ability to handle complex clinical situations arising in illnesses with multiple etiologies, to evaluate the physiological validity of diagnostic possibilities being explored, and to organize large amounts of seemingly unrelated facts into coherent causal descriptions.

1. Introduction

We have studied difficulties arising in the operation of the "first generation" of AI programs in medicine and have undertaken the development

Reprinted from the *Proceedings of the Seventh International Joint Conference* on Artificial Intelligence, pp. 893–899, 1981. Used by permission of the International Joint Conferences on Artificial Intelligence, Inc.; copies of the Proceedings are available from William Kaufmann, Inc., 95 First Street, Los Altos, CA 94022, U.S.A.

of knowledge representation structures to support needed improvements. The description of a patient in existing programs such as INTERNIST-1 [4], PIP [3], and MYCIN [6] starts from a single list of findings about the patient. Using a database of associations between diseases and findings (or rules establishing those connections), these programs form an interpretation of the patient's condition which is essentially a list of possible diseases, ranked by a calculated estimate of likelihood or degree of belief in each.

Researchers have recognized the need [2,5,7] to use notions such as causal relationships, temporal patterns, and aggregate disease categories in the description of a program's diagnostic understanding, but the mechanisms provided to do this have been too weak. For example, although causality appears as a term in descriptions in PIP and INTERNIST-1, in both cases its use is limited to guiding the propagation of likelihood measures. These programs fail to capture the human notion that explanation should rest on a chain of cause-effect deduction. Although, the CASNET/ Glaucoma [10] program uses a network of causally-related states and defines diseases as paths in this network, its primary reasoning mechanism is nevertheless the local propagation of probability weights.

Similarly, it has been realized that a diagnostic or therapeutic program must consider a case at various levels of detail in order to integrate its overall understanding with its detailed knowledge. This insight also has not prevailed in the actual mechanisms provided in existing programs.

To explore the issues outlined here, we have undertaken a study of the medical problem of providing expert consultation in cases of electrolyte and acid-base disturbances. We have partly completed implementation of a program, ABEL, which is the diagnostic component of our overall effort. In this paper we concentrate on ABEL's mechanism for describing a patient. Called the patient-specific model (PSM) [1], this description includes data about the patient as well as the program's hypothetical interpretations of these data in causal hierarchical networks. We describe the representations of medical knowledge and the processing strategies needed to enable ABEL to construct a PSM from the initial data presented to the program about a patient. The same representations and procedures will also be useful to revise the PSM during the process of diagnosis, but we will concentrate here on the logically prior operation of building the PSM.

Our understanding of medical expert reasoning suggests that an expert physician may have an understanding of a difficult case in terms of several levels of details. At the shallowest that understanding may be in terms of commonly occurring associations of syndromes and diseases, whereas at the deepest it may include a biochemical and pathophysiological interpretation of abnormal findings. For our program to reason at a sophisticated level of competence, it will need to share such a range of representations. The PSM is, therefore, a multi-level causal model, each level of which attempts to give a coherent account of the patient's case. This model also serves as the basis for an English generation facility that provides explanations of the program's understanding.

The PSM is created by instantiating portions of ABEL's general medical knowledge and filling in its details from the specific case being considered. The instantiation of the PSM is very strongly guided by initially given data, because the PSM includes only those disorders and connections that are needed to explain the current case. Instantiation is accomplished by five major operators. *Aggregation* and *elaboration* make connections across the levels of detail in the PSM by filling in the structure above and below a selected part of the network, respectively. In a domain such as ABEL's, multiple disorders in a single patient and the presence of homeostatic mechanisms requires the program to reason about the joint effects of several mechanisms which collectively influence a single quantity or state. *Component decomposition* and *summation* relate disorders at the same level of detail by mutually constraining a total phenomenon and its components; the net change in any quantity must be consistent with the sum of individual changes in its parts. The final operator, *projection,* forges the causal links within a single level of detail in the search for etiologic explanations. The operators all interact because the complete PSM must be self-consistent both within each level and across all its levels. Therefore, each operation typically requires the invocation of others to complete or verify the creation of related parts of the PSM.

2. Hierarchical Representation of Medical Knowledge

Based on our observation that a physician's knowledge is expressed at various levels of detail, we have developed a hierarchical multi-level representation scheme to describe medical knowledge and procedures to instantiate this knowledge to describe a particular patient's illness. The lowest level of description consists of pathophysiological knowledge about diseases, which is successively aggregated into higher level concepts and relations, gradually shifting the content of the description from physiological to syndromic knowledge. The aggregate syndromic knowledge provides us with a concise global perspective and helps in the efficient exploration of diagnostic possibilities. The physiological knowledge provides us the capability of handling complex clinical situations arising in patients with multiple disturbances, evaluating the physiological validity of the diagnostic possibilities being explored, organizing large number of seemingly unrelated facts and formulating therapy recommendations and prognosis. Finally, as the causal-physiological reasoning tends to be categorical and syndromic reasoning probabilistic, the hierarchical description allows us to blend together the use of categorical and probabilistic reasoning.[8]

2.1 Multi-Level Description of States

Medical knowledge about different diseases and their pathophysiology is understood to varying degrees of detail. While it may be easier for a program to reason succinctly with medical knowledge artificially represented at a uniform level of detail, we must be able to reason with medical knowledge at different levels of detail to exploit all the medical information available. Although this does not pose any difficulty in medical domains where the pathophysiology of diseases is not well developed, in a domain such as electrolyte and acid-base disturbances where, on the one hand, the pathophysiology of the disturbances is well developed, and on the other, the pathophysiology of many of the diseases leading to these disturbances is relatively poorly understood, we are constantly faced with this problem.

Secondly, the information about a patient parallels the physician's medical knowledge about diseases and therefore also comes at different levels of detail. For example, "serum creatine concentration of 1.5" is at a distinctly different level than "high serum creatine,"[1] and "lower gastrointestinal loss" than "diarrhea." We need some mechanism by which we can interrelate these concepts. Finally, in order to be effective in diagnostic problem solving and communicating with clinicians we ought to have the ability to portray the diagnostic problem in a small and compact space. Yet to be efficacious, we must maintain the ability to take every possible detail into consideration. We have solved this problem by representing the medical knowledge in five distinct levels of detail from a deep pathophysiological level to a more aggregate level of clinical knowledge about disease associations.

Each level of the description can be viewed as a semantic net describing a network of relations between diseases and findings. Each node represents a normal or abnormal physiological state and each link represents some relation (causal, associational, etc.) between different states. A state (interchangeably used with node) in the system such as "diarrhea" is represented as a node in the causal network. Each node is associated with a set of attributes describing its temporal characteristics, severity or value, and other relevant attributes. A state is called a *primitive-node* if it does not contain internal structure and is called a *composite-node* if it can be defined in terms of a causal network of states at the next more detailed level of description. One of the nodes in this causal network is designated as the *focus node* and the causal network is called the *elaboration structure* of the composite node. The focus node identifies the essential part of the causal structure of the node above it. Indeed, the collection of focal nodes acts to align the causal networks represented by different levels of the PSM. We note that very often a composite node and its focal descrip-

---

[1] For a muscular patient whose previously known value of creatinine is 1.3 we can assume this to be normal, but for a patient with a previously known value of 1.0 this is definitely high and could imply a loss of about ⅓ of the kidney function.

	Lower-GI-Fluid	Plasma-Fluid
Na	100–110	138–145
K	30–40	4–5
Cl	60–90	100–110
HCO3	30–60	24–28

Figure 1.

tion at the next level share the same name; this is typical in English, where the level of detail of place names, for example, is often obtained from context and not encoded in the name used. Nodes that do not play a role as the focal definition of any node at a higher level are called *non-aggregable nodes*. They represent a detailed aspect of the causal model which is subsumed under other nodes with different foci at less detailed levels of description.

To illustrate the description of a state at various levels of aggregation, let us consider the electrolyte and acid-base disturbances that occur with diarrhea, which is the excessive loss of lower gastrointestinal fluid (lower GI loss). The composition of the lower gastrointestinal fluid and plasma fluid are as follows (see Figure 1). In comparison with plasma fluid, the lower GI fluid is rich in bicarbonate (HCO_3) and potassium (K) and is deficient in sodium (Na) and chloride (Cl). This information is represented in the knowledge base by decomposing lower GI loss into its constituents (and associating appropriate quantitative information with the decomposition). The loss of lower GI fluid would result in the loss of corresponding quantities of its constituents (in proportion to the total quantity of fluid loss) as shown in Figure 2. Therefore, an excessive loss of lower GI fluid without proper replacement of fluid and electrolytes would result in a net reduction in the total quantity of fluid in extracellular compartment (hypovolemia). Because the concentration of K and HCO_3 in lower GI fluid is greater than in plasma fluid, there is a corresponding reduction in the concentration of K (hypokalemia) and HCO_3 (hypobicarbonatemia) in the extracellular fluid. Finally, as the concentration of Cl and Na in the lower GI fluid is lower than that in plasma fluid, there is an increase in the concentration of Cl (hyperchloremia) and Na (hypernatremia) in the extracellular fluid. This is represented at the next

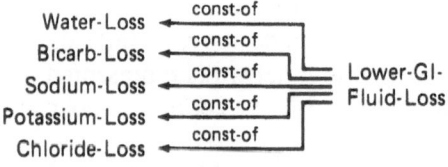

Figure 2.

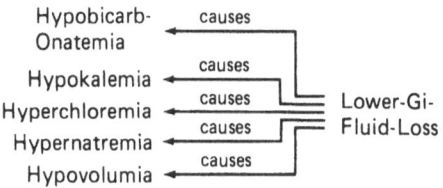

Figure 3.

level of description as shown in Figure 3. Figure 4 shows the aggregation of this information along with some additional causes and consequences of lower GI loss at the next more aggregate level of detail. The lower GI loss at this level is a non-aggregable state and therefore does not have an aggregation at the next level above. Figure 5 shows the description of the aggregate effects of diarrhea (one of the causes of lower GI loss). The summarization of the description of lower GI loss and diarrhea shown in Figure 5 is achieved through the use of link aggregation and elaboration, described in the next subsection.

2.2 Multi-Level Description of Causal Links

A causal link specifies the cause-effect relation between the cause (the antecedent) and the effect (the consequent) states. In past programs (e.g., PIP, INTERNIST), causal links were described by specifying the type of causality (may-be-caused-by, complication-of, etc.), and a number or a set of numbers representing in some form the likelihood (conditional probability), importance, etc., of observing the effect given the cause or vice versa. We now believe that this simple representation of the relation between states is inadequate. The form of presentation of an effect and the conditional probability of observing it depends upon various aspects of the cause such as severity, duration, etc., as well as other factors in the context in which the link is invoked[2] (such as the patient's age, sex and weight, and the current hypothesis about the patient). Therefore a causal

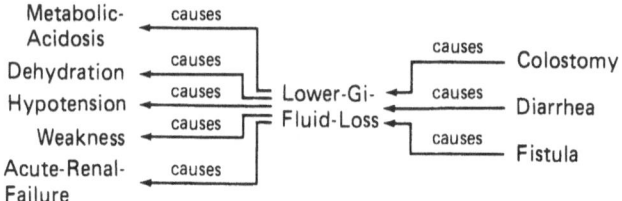

Figure 4.

[2]For example, a severe diarrhea causes severe hypokalemia, and a mild diarrhea causes mild hypokalemia.

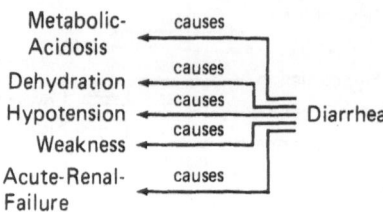

Figure 5.

link in the system (an object denoting the causal relation between a cause-effect pair) specifies a multivariate relation between various aspects of the cause and effect, and also specifies the context and assumptions which constrain the causal relation, as shown in Figure 6.

One important function of diagnostic reasoning is to relate causally the diseases and symptoms observed in a patient. These causal relations play a central role in identifying clusters that can be meaningfully aggregated in developing coherent diagnoses. The presence or absence of a causal relation between a pair of states can change their diagnostic and prognostic interpretations. Therefore, the system should and does have the capability of hypothesizing the presence or absence of a causal link. This is the reason why links are objects in their own right rather than simple pointers between nodes.

To reason with a causal network representation effectively, a program must make conclusions about a node or link depending only on information that is locally available from the neighborhood of the mechanism in question. If nonlocal effects are to be invoked in causal explanations, they must be explicitly identified (e.g., as part of the context of the causal link) or else they corrupt our ability to reason with any portion of the network. If at some level of detail two distant phenomena interact, we must aggregate the description of the causal network to a level where the two phenomena are adjacent to one another. Further, because the causal relations specified by links are not guaranteed to be true under all circumstances (they represent strong associations, not logical truth), the validity of deductions degrades with every additional intermediate link. That is,

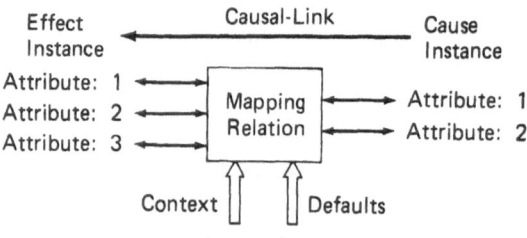

Figure 6.

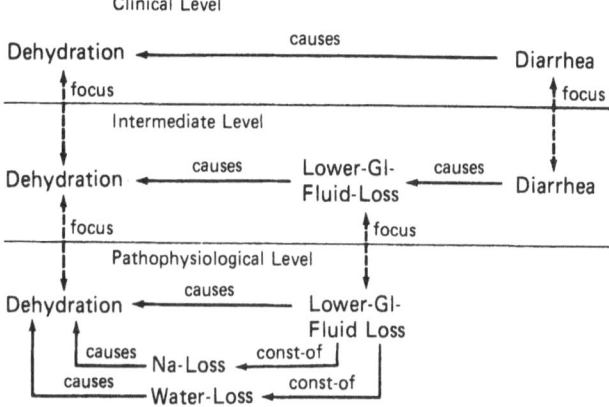

Figure 7.

a causal pathway containing a large number of links is less likely to be valid than one using only a few links. Therefore, in order to explore a large diagnostic space, we must aggregate the diagnostic space to a level where each link represents an aggregate causal phenomenon covering larger distances and thus minimizing the possibility of error in the deduction. This ability to move from one level of description to another is provided by the multi-level description proposed here.

Links can be categorized, as nodes are, into two types: the *primitive links* and the *composite links*. To illustrate the concept of elaborating causal links to form a causal pathway, let us consider the causal relation between diarrhea and dehydration shown in Figure 7. The causal mechanism of diarrheal-dehydration can be elaborated as follows: diarrhea causes lower GI loss, which causes dehydration. Expressed at the next level of detail, the lower GI fluid loss can be described as consisting of the loss of water and sodium along with other electrolytes. The water loss in the presence of the reduced total quantity of extracellular sodium results in lower extracellular volume, which at the higher level of description is described as dehydration.

3. Reasoning About Components

One of the important areas of medical diagnosis not adequately addressed by the first generation of AIM programs is the evaluation of the effect of more than one disease present in the patient simultaneously, especially when one of the diseases alters the presentation of the others. This problem does not place serious limitations on programs dealing with single problems such as the therapy of glaucoma or the diagnosis of baceremia. But, in the case of electrolyte and acid-base disturbances where a large fraction of cases involve multiple diagnoses, the ability to evaluate the joint influence of multiple disease and the ability to decompose their influences on observable findings is particularly important.

For example, let us consider a patient with diarrhea and vomiting leading to severe hypokalemia. Let us also suppose that we know about the diarrhea, but we are not aware of the vomiting. The observed hypokalemia is too severe to be properly accounted for by the diarrhea alone. Without the ability to decompose the hypokalemia, we would have to attribute it completely to the diarrhea or completely to something else. In either case[3] we fail because the total state of hypokalemia is inconsistent with any of its possible single causes. Thus, any single cause hypothesized by the program (e.g., vomiting) will not be severe enough to account for the observed hypokalemia by itself. As argued above, we need the ability to hypothesize that only a part of the hypokalemia is accounted for by diarrhea. We introduce the notion that any primitive node in the causal hierarchy[4] may have *components,* which are other primitive nodes which together make up the given node.

In our system this is achieved by a pair of operators: *component summation* and its dual *component decomposition.* Using our example, these operators allow us to attribute only a part of hypokalemia to the diarrhea and to compute that part of hypokalemia that is not caused by diarrhea (called the *unaccounted component* of the hypokalemia). These operations deal not only with the magnitude of some disorder but also with other attributes such as duration. They are implemented by associating with each primitive node a multivariate relation that constrains attributes of the node and its constituents. Component summation combines attributes of the components to generate the attributes of the joint node: component decomposition identifies unaccounted components by noting differences between the joint node and its existing components. These operations enrich the PSM by instantiating and unifying together component nodes when the case demands them. This occurs whenever multiple causes contribute jointly to a single effect. An important case of this arises whenever feedback is modeled, because in any feedback loop there is at least one node acted on both by an outside factor and by the feedback loop itself.

As the PSM is built, component summation and decomposition operations can cause a node in the program's general knowledge to be instantiated as a node and its several components. If a node is primitive and

---

[3] All the previous programs would allow the entire hypokalemia to be accounted for by diarrhea. In particular, PIP after allowing the hypokalemia to be accounted for by diarrhea will not allow hypokalemia to lend any support to the hypothesis of vomiting. INTERNIST-1, on the other hand, will allow the entire hypokalemia to lend support to the hypothesis of vomiting as well as allowing it to be explained by diarrhea.

[4] Recall that primitive means that it is not the aggregation of a further-defined causal structure.

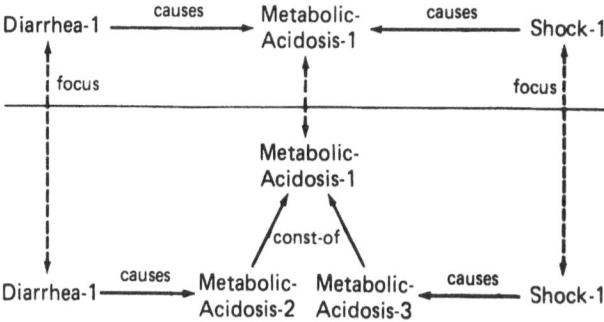

Figure 8.

there are multiple causes, the contribution of each cause is instantiated separately. Then the profile of the combination is computed using component summation. The combined effect is then instantiated and connected to its constituents by constituent links.

Because components are defined only for primitive nodes, the instantiation of composite nodes which involve component summation must be in terms of the summation of components in the node's elaboration structure. If the node is composite then we elaborate the constituent nodes around their focal nodes until we reach the primitive nodes associated with them at a level of greater detail. Then we combine these primitive nodes and aggregate their effects back. For example, if we know that a patient has two disturbances, diarrhea and shock, causing metabolic-acidosis (Figure 8), then we evaluate their contribution to metabolic-acidosis and then focally elaborate the two constituents until the metabolic-acidosis is described in terms of the quantity of serum bicarbonate lost.[5] We then aggregate the joint effects to derive the actual severity of metabolic-acidosis.

As mentioned above, the mechanism of component summation allows us to represent feedback explicitly by representing the primary component of the change (the forward path) and the secondary feedback component (the response of the homeostatic mechanism in defense of the parameter being changed) as components to be summed to yield the whole. Figure 9[6] shows the primary change in serum pH caused by low serum bicarbonate and the response of the respiratory system in defense against the change in serum pH. Read the example as follows: the lowering of the concentration of serum bicarbonate causes a reduction in serum pH which causes hyperventilation and thus reduces the pCO_2,

---

[5]The overall quantity of serum bicarbonate lost may be computed simply by adding the loss due to each of its causes.

[6]This is a hypothetical example; in the program this component summation will take place at the pathophysiological level.

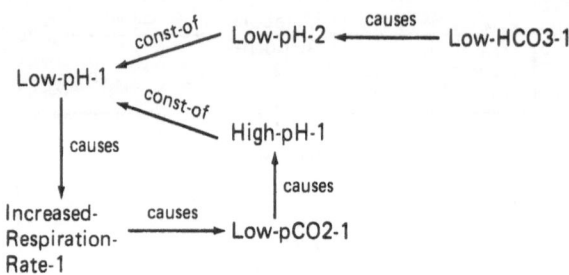

Figure 9.

which in turn causes an increase in the serum-pH (negative feedback). This increase is less than the initial reduction, causing a net reduction in serum pH. The decomposition of an effect with multiple causes into its causal components also provides us with valuable information in evaluating prognosis and in formulating therapeutic interventions.

4. The Patient Specific Model

Diagnosis is the process of actively seeking information and identifying the disease process(es) causing the patient's illness. In other words, diagnosis involves ascertaining the facts and their implications. The effectiveness of the information gathering process depends on the analysis of the available facts. From our experience with the existing diagnostic systems [3,4], we are convinced that a relatively simple representation of physician's analysis of patient's illness (i.e., a list of disease hypotheses) is incapable of providing the desired level of expertise. The patient description must unify all the known facts about the patient, their interpretations, their suspected interrelationships and disease hypotheses in order to explain these findings. Finally, we observe that at any point in diagnostic reasoning practiced by human experts, there are only a few significantly different explanations for the patient's illness under consideration.

In the program, each such explanation is represented by a patient specific model (PSM). Note that within each PSM all the diseases, findings, etc., are mutually complementary, while the alternate PSM's are mutually exclusive and competing. In this section we describe procedures for building and extending a patient specific model based on the known findings and the program's medical knowledge. These operations are *initial formulation* to create an initial patient description from the presenting complaints and laboratory results, *aggregation* to summarize the description at a given level of detail to the next more aggregate level, *elaboration* to elaborate the description at a given level of aggregation to the next more detailed level and *projection* to hypothesize associated findings and diseases suggested by states in the PSM.

4.1 Initial Formulation

From observing the clinical behavior of physicians, we have noticed that when presented with the chief complaints and other voluntarily provided information in a case, the physicians set up a tentative diagnosis. This diagnosis serves as a specific framework which can be used in soliciting information and for organizing the incoming information. Similarly the program when provided with the initial findings and a set of serum electrolyte values constructs a small set of PSM's as its initial possible diagnoses, using the following steps. First it analyses the electrolytes and formulates all possible single or multiple acid-base disturbances that are consistent with the electrolyte values provided and selects from them a small set which is consistent with the initial findings. Next, it generates a pathophysiological explanation of the electrolytes based on each of the proposed acid-base disturbances. This is performed by elaborating all known clinical information to the pathophysiological level, where its relationships to the laboratory data are determined by projecting the unique causes and definite consequences of every node. Then the program summarizes these pathophysiological descriptions to the clinical level by repeated application of aggregation operations. This process results in the initial description of the patient at every level of detail. It is this description which is later modified by the diagnostic process as new information becomes available. Note that each of the mechanisms, aggregation, elaboration and projection, are used in the initial formulation of the PSM.

4.2 Aggregation

The aggregation process allows us to summarize the description of the patient's illness at any given level to the next more aggregate level. The summarization of a causal network can be achieved by recognizing that a central node and its surrounding causal relationships may be expressed at a more aggregate level by a single node (called *focal aggregation*), and by summarizing a chain of relations between nodes by a single causal relation between the initial cause and the final effect nodes (called *causal aggregation.*).

4.2.1 Focal Aggregation. In aggregating a causal network we must first identify the nodes in the network that form anchor points (i.e., landmarks, points of special significance) around which the causal phenomenon can be summarized. Consider a partially-completed PSM in which some nodes at a detailed level of aggregation have been instantiated. Any of these nodes is an anchor point if: (1) in the medical knowledge base such a node is the focus of some node at the next more aggregate level in the network, and (2) at least one such higher level node already exists or can be instantiated within the PSM. If it exists and the constraints on the focal link are satisfied, then the focal link connecting the two is instantiated. If it does not exist, then both it and the focal link are instantiated. Finally, if more than one possible description of the node is consistent with the causal structure above, we defer the aggregation process until we can obtain some additional information to resolve this ambiguity.

4.2.2 Causal Aggregation. Once we have determined the focal aggregations for nodes at a given level of aggregation we need to describe the causal relations among these aggregate nodes. The process of causal aggregation takes a node and its causes and aggregates the relation between them according to one of three rules. First, if the node has no causal predecessors or if none of the causal paths leading into the node (called *predecessor paths*) have a node with a focal aggregation then the focal aggregation of the node either is an ultimate etiology or is totally unaccounted for and does not need to be causally aggregated. Second, if every predecessor path has a node with a focal aggregation then the focal aggregation of the node is fully accounted for. The causal aggregation is achieved by instantiating a causal link between the focal aggregation of the node and the first focal aggregation in each path. Finally, if only some of the predecessor paths have nodes with focal aggregations then the focal aggregation of this node is partially accounted for. The causal aggregation is achieved by decomposing the node into two components: (1) due to paths which have focal aggregation (called *accounted component*): and (2) due to paths that do not (called *unaccounted component*). Then, the focal aggregation of the node is decomposed based on the decomposition at the present level and the two cases are treated as described above.

4.3 Elaboration

Elaboration is the dual of the aggregation operation described above and is used to elaborate the description of a causal network at a given level of aggregation to the next more detailed level. This is achieved by elaborating each link in the causal network by first describing the cause and effect of the link at the next more detailed level (called *focal elaboration*) and then instantiating the causal pathway between these detailed nodes (called *causal elaboration*). If the causal pathway being instantiated interacts with other causal paths in the PSM, the combined effects of the multiple causality are computed using component summation. The combined effects of this summation can then be aggregated to reflect the better understanding of the causal phenomenon at higher levels of aggregation.

4.3.1 Focal Elaboration. Focal elaboration is the inverse of focal aggregation. To focally elaborate a composite node the program computes the possible profile of the focal concept associated with the given node. If a node at the next lower level of aggregation matches this profile and is consistent with the node above, it instantiates the focal link connecting the two. If not, it instantiates the focal node and the focal link connecting the two.

4.3.2 Causal Elaboration. Causal elaboration is the dual of causal aggregation. A composite causal link can be elaborated if the cause and the effect nodes of the link have focal elaborations. To elaborate a composite link the program matches the causal path associated with the link starting at the focal nodes of the cause and the effect of the link with existing paths in the PSM. If some part of this pathway is not present in the PSM, the program recursively calls itself on each link in the pathway (starting from the focus node of the source) that is absent in the PSM. If the link being recursively elaborated is a primitive link and if its effect node is not present in the PSM, the effect node and the link are instantiated. Otherwise, if the effect node is present, it matches the attributes of the cause and the effect nodes. If they are compatible, it instantiates the link. Otherwise, if the effect node is an observed node,[7] the program decomposes the effect node and instantiates the link connecting the cause and the component of the effect node contributed by it. Otherwise, if the effect node is accounted for by some other cause, it instantiates the combined effect by summing the components of the two causes. Finally, it aggregates the effect node to revise the description at the next more aggregate level.

4.4 Projection

The Projection operation is used to hypothesize and explain the associated findings and diseases suggested by the states in the PSM. The projection operation is very similar to elaboration. It differs from elaboration in that the causal relation being projected is hypothetical and therefore is not present in the PSM. Furthermore, the projection operation fails if the causal description of the hypothesized link is inconsistent with the description in the PSM at any level of aggregation. As a result, the application of the projection operation cannot result in the decomposition of a fully accounted node, creating an additional unaccounted component and therefore degrading the quality of explanation.

We envision using the projection operation in the diagnostic problem solver for exploring diagnostic possibilities, evaluating their physiological validity and in generating expectations about the consequences of hypothesized diagnoses.

5. An Example

Let us consider a 40 year old 70 Kg patient who has been suffering from moderately severe diarrhea for the last two days and, as a result, had

---

[7]or if the effect node is a causal predecessor of some observed node that completely accounts for it.

developed moderately severe metabolic acidosis and hypokalemia. The laboratory analysis of the patient's blood sample (serum analysis) is; Na: 140, K: 3.0, Cl: 115, HCO_3: 15, pCO_2: 30, and pH: 7.32.

5.1 Initial Formulation

To exercise the program, let us provide it initially with only the laboratory data. Based on these data, the program generates all possible acid-base disturbances that can account for the laboratory data. They are:

1. metabolic-acidosis
2. chronic-respiratory-alkalosis + acute-respiratory-acidosis
3. metabolic-acidosis + chronic-respiratory-alkalosis
   + acute-respiratory-acidosis
4. metabolic-alkalosis + chronic-respiratory-alkalosis
   + acute-respiratory-acidosis.

Based on the complexity,[8] likelihood and severity of each component, the list of possible disturbances is pruned and rank-ordered.[9] The rank-ordered list of likely disturbances is

1. metabolic-acidosis (severity: 0.4)
2. chronic-respiratory-alkalosis (severity: 0.68)
   + acute-respiratory-acidosis (severity: 0.32).

The program now creates a PSM[10] for each possible acid-base disturbance and asserts in it instantiations of the laboratory data (at the pathophysiological level) and the appropriate acid-base disturbances (at the clinical level). In the rest of the example we will focus on the first acid-base disturbance, namely metabolic acidosis. The program focally elaborates the metabolic acidosis through the intermediate levels until it reaches the pathophysiological level and thus identifies the amount of HCO_3 loss corresponding to the severity of the metabolic-acidosis. Based on this information and the laboratory data, it instantiates the feedback loop corresponding to the acid-base homeostatic mechanism. Next, it projects back[11] each node whose cause can be uniquely determined and projects forward the definite consequences of each node in the PSM. We now have the pathophysiological level explanation of the electrolytes consistent with the diagnosis of metabolic-acidosis as shown in Figure 10.

[8]Triple disturbances are quite rare and are generally not considered during initial formulation unless there is compelling evidence for their presence.

[9]The rank ordering of the diseases is based on Occam's Razor—simpler hypotheses are preferred.

[10]For ease of explanation, the example described here uses a three level PSM instead of the five level PSM used in the program.

[11]Note here that as we are at pathophysiological level, each link being projected is primitive. Thus, projecting back at this level can be restated as instantiating the cause and the link connecting the cause and the effect node.

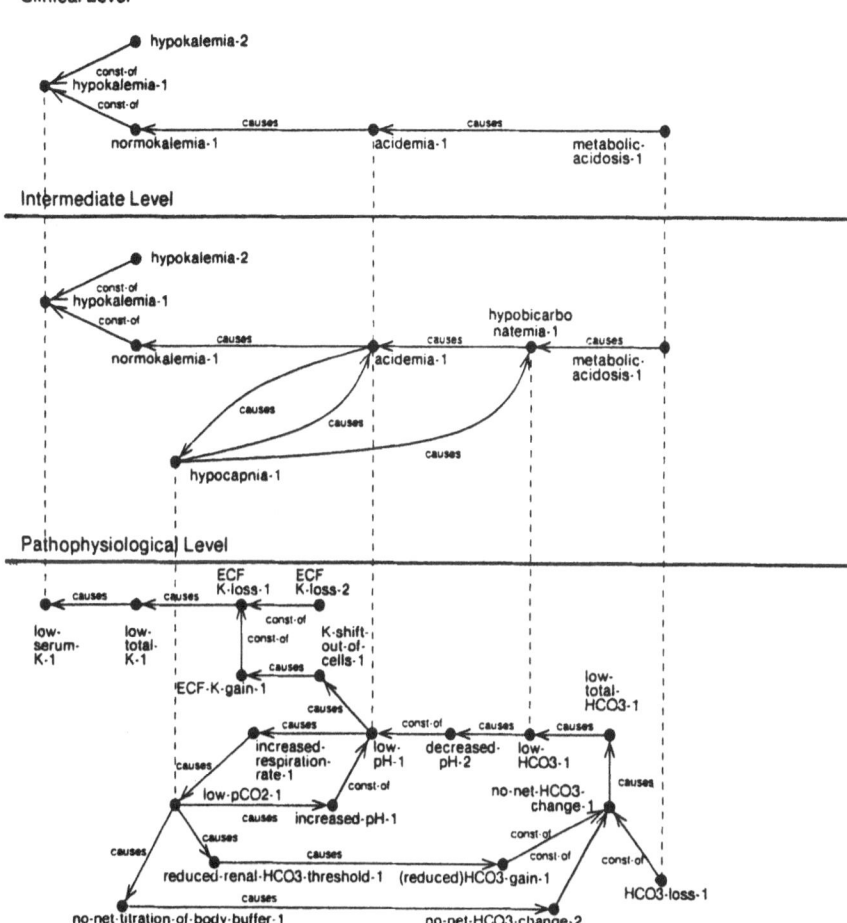

Figure 10. Initially formulated PSM.

5.2 Aggregation

After the pathophysiological description is completed, this description is aggregated through the intermediate levels to the clinical level of detail. To illustrate this operation let us consider the low-serum-K-1 node at the pathophysiological level. Focally aggregating this node, we instantiate hypokalemia-1 as shown in Figure 10. To determine the causal aggregation of this node at the next level of detail we must focally aggregate the first aggregable node in each path leading back; in this case low-pH-1. Focally aggregating low-pH-1 we instantiate acidemia-1. Next, we compute the component of low-serum-K that can be accounted for by low-pH-1 and the component that remains to be accounted for because of the unaccounted K-loss-2. Then we compute the mapping of these components at the next level of aggregation and instantiate normokalemia-1 (the component accounted by low-pH-1) and hypokalemia-2 (due to unaccounted K-loss-2). We then connect the normokalemia-1 to acidemia-1 and mark the hypokalemia-2 as unaccounted (indicated in the figure by an asterisk). Next, in order to causally aggregate low-pH-1 we focally aggregate low-pCO_2-1 and low-HCO_3-1 into hypocapnia-1 and hypobicarbonatemia-1 respectively. As each path leading back from low-pH-1 terminates in a node with focal aggregation, the focal aggregation of low-pH-1 (acidemia-1) is a fully accounted node. Therefore, we connect acidemia-1 to hypocapnia-1 and hypobicarbonatemia-1. This process is repeated for each aggregable node at the current level and then the whole process is repeated at the next level until we reach the clinical level of aggregation.

5.3 Projection

To illustrate the projection operator, let us assume that the diagnostic component has hypothesized that the unaccounted component of hypokalemia at the clinical level (hypokalemia-2) is caused by diarrhea and wishes to determine if this is so and how this assumption fits with the current PSM. The result of this operation is shown in Figure 11.

To project the link between hypokalemia and diarrhea the program evaluates the link to determine the attribute profile of the diarrhea consistent with hypokalemia-2 from which it determines the profile of diarrhea at the next more detailed level. It then attempts to match the causal path associated with the link (hypokalemia ← lower-GI-loss ← diarrhea) at the next level. As none of the links in this pathway are present and as this causal pathway is consistent with the description at the next level, the program recursively calls itself on each link in the path. Considering the first link (that is, hypokalemia ← lower-GI-loss), it finds the causal path associated with this link at the next level of detail (low-serum-K ← low-total-K ← K-loss ← lower-GI-loss). Matching this path with the description in the PSM, it finds that all but one link (K-loss ← lower-GI-loss) is already present. Since this link is primitive, the program evaluates the profile of the lower-GI-loss consistent with the unaccounted component of K-loss and instantiates it and the causal link connecting lower-GI-loss-1 to K-loss-2. To reflect this addition at the higher levels of detail, the program aggregates the low-serum-K-1 (the effect node in the path). As the low-serum-K-1 is now a fully accounted node, the component structure associated with its focal aggregation (hypokalemia-1) is deleted and the causal links associated with the accounted component of hypokalemia-1 and an additional link from lower-GI-loss-1 are connected to it. This process is repeated until we establish the relation between the diarrhea and hypokalemia at the clinical level.

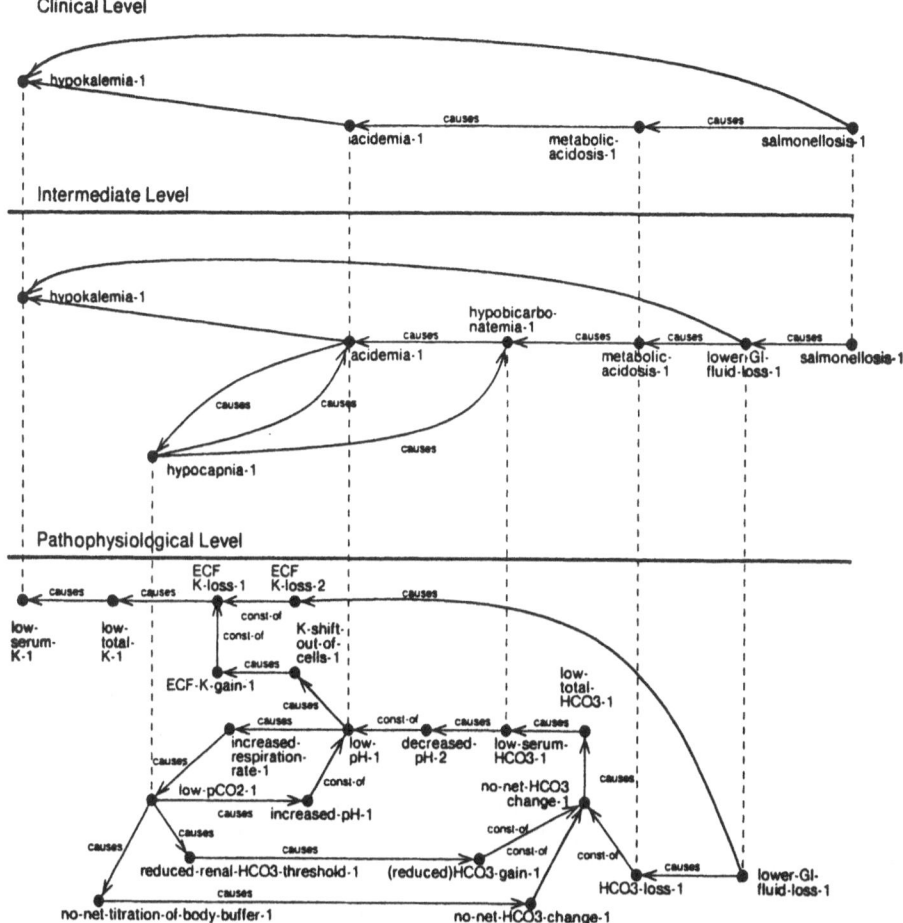

Figure 11. PSM extended to include diarrhea.

5.4 Elaboration

The process of elaboration is similar to that of projection described above and differs from it in two major ways; (1) the causal link and the associated nodes already exist in the PSM at the higher level of aggregation and (2) we have already determined that the causal link being elaborated is valid. Therefore if a causal pathway associated with the link at some level of detail is not consistent with the description in the PSM, the program modifies the PSM appropriately to accommodate the pathway. In the example being described the second (and more interesting) case does not arise. To demonstrate the elaboration process let us establish the relation between diarrhea-1 and metabolic-acidosis-1 at the clinical level. The result of elaborating this link is shown in Figure 11.

Clinical Level

This is a 40 year old 70.0 kg male patient with moderate diarrhea. His electrolytes are:

Na: 140.0 HCO_3: 15.0   Agap: 13.0
K: 3.0 pCO_2: 30.0
Cl: 115.0   pH: 7.32

The diarrhea causes moderate metabolic acidosis, which causes mild acidemia. The acidemia and diarrhea cause mild hypokalemia and acidemia causes hyperventilation. All findings have been accounted for.

Intermediate Level

This is a 40 year old 70.0 kg male patient with moderate diarrhea. His electrolytes are:. . . .

The diarrhea causes moderate lower GI loss, which causes moderate metabolic acidosis. The metabolic acidosis along with moderate hypocapnia causes moderate hypobicarbonatemia. The hypobicarbonatemia along with hypocapnia causes mild acidemia. The acidemia and lower GI loss cause mild hypokalemia and acidemia causes hypocapnia. The acidemia also causes hyperventilation. All findings have been accounted for.

Pathophysiological Level

This is a 40 year old 70.0 kg male patient with moderate lower GI loss. His electrolytes are:. . . .

Moderate lower GI loss, reduced renal HCO_3 threshold and normal HCO_3 buffer binding jointly cause no HCO_3 change. The no HCO_3 change causes low ecf HCO_3, which causes low serum HCO_3. The low serum HCO_3 and low serum pCO_2 jointly cause low serum pH. The low serum pH causes K shift out of cells and causes increased respiration rate. The increased respiration rate causes low serum pCO_2, which causes normal HCO_3 buffer binding. The low serum pCO_2 also causes reduced renal HCO_3 threshold and increased respiration rate causes increased ventilation. The lower GI loss and K shift out of cells jointly cause K loss. The K loss causes low ecf K, which causes low serum K. All findings have been accounted for.

Figure 12. English explanation at different levels of detail.

6. English Explanation

To illustrate the program's understanding of the patient's illness at various levels of detail, an English generator was implemented to translate the PSM at any given level into its English description.[12] The descriptions are given at three levels of detail in Figure 12.

---

[12]The generator makes use of the methodology and some of the code of a generator built by William Swartout as part of an interactive system which explains and justifies portions of expert programs [9].

7. Conclusion

We have begun a complex and challenging task: to reason about difficult medical problems with a representation that is capable of capturing the subtlety and richness of knowledge and hypotheses used by expert physicians. We have thus far succeeded in creating a representation and a set of structure building operators which are able to create a patient description based on causal models, multiple levels of detail in description, and the explicit use of components of quantities and states. The various viewpoints on the patient respresented by different cuts through this complex description are kept consistent by the operators. We believe that this approach displays a level of understanding not achieved before in medical reasoning programs or others which need to describe an organization of hypotheses or mechanisms at different levels of detail.

In continuing to develop our diagnostic and therapeutic programs, we believe that the organizational framework provided by the PSM and its associated operators gives us a suitable machinery for exploring the choice of reasoning strategies and recording our programs' changing conceptions of a case. The rich network of interconnections in the PSM constrains a diagnostic reasoner to generate only a relatively small number of coherent explanations, thereby reducing the space of possibilities to be investigated in seeking a diagnosis. In particular, enforcing the requirements of causal consistency (at each appropriate level of detail) on any tenable explanation provides a means of pruning the diagnostic space and permits us to try a "hypothesize and debug" reasoning strategy. The multiple level interconnections of the PSM also help us merge decisions and considerations we have described as categorical and probabilistic. Although much work clearly remains before developments such as those described here form the fabric of truly successful medical consulting systems, we have proposed here a useful new representational basis for such work.

Acknowledgments. This research was supported (in part) by the National Institutes of Health Grant No. 1 PO1 LM 03374-02 from the National Library of Medicine.

References

[1] Gorry, G.A., Silverman, H., and Pauker, S.G., Capturing Clinical Expertise: A Computer Program that Considers Clinical Responses to Digitalis, *American Journal of Medicine* 64:452–460 (March 1978).
[2] Patil, R.S., Design of a Program for Expert Diagnosis of Acid Base and Electrolyte Disturbances, MIT Laboratory for Computer Science TM-132 (May, 1979).

[3] Pauker, S.G., Gorry, G.A., and Kaissirer, J.P., and Schwartz, W.B., Toward the Simulation of Clinical Cognition: Taking a Present Illness by Computer, *The American Journal of Medicine* 60:981–995 (June 1976).

[4] Pople, H.E., Jr., Myers, J.D., and Miller, R.A., DIALOG: A Model of Diagnostic Logic for Internal Medicine, *Advance Papers of the Fourth International Joint Conference on Artificial Intelligence,* available from the Artificial Intelligence Laboratory, Massachusetts Institute of Technology (1975).

[5] Pople, H.E., Jr., The Formation of Composite Hypotheses in Diagnostic Problem Solving: an Exercise in Synthetic Reasoning, *Proceedings of the Fifth International Joint Conference on Artificial Intelligence,* available from the Department of Computer Science, Carnegie-Mellon University. Pittsburgh, PA 15213. (1977).

[6] Shortliffe, E.H., *Computer Based Medical Consultations: MYCIN,* Elsevier North Holland Inc. (1976).

[7] Smith, B.C., A Proposal for a Computational Model of Anatomical and Physiological Reasoning. Artificial Intelligence Laboratory, Massachusetts Institute of Technology, AI-Memo 493 (1978).

[8] Szolovits, P., Pauker, S.G., Categorical and Probabilistic Reasoning in Medical Diagnosis, *Artificial Intelligence* 11:115–144 (1978).

[9] Swartout W.R., Producing Explanations and Justifications for Expert Consulting Pipgrams, Laboratory for Computer Science, Massachusetts Institute of Technology, Technical Report LCS-TR-251 (1981).

[10] Weiss, S.M., Kulikowski, C.A., Amarel, S., and Safir, A., A Model-Based Method for Computer-Aided Medical Decision-Making, *Artificial Intelligence* 11:145–172 (1978).

Index